WHY BOB DYLAN MATTERS

WHY BOB DYLAN MATTERS

RICHARD F. THOMAS

DEY ST.
An Imprint of WILLIAM MORROW

HarperCollins books may be purchased for educational, business, or sales promotional use. For information, please email the Special Markets Department at SPsales@harpercollins.com.

A hardcover edition of this book was published in 2017 by Dey Street, an imprint of William Morrow.

FIRST DEY STREET PAPERBACK EDITION PUBLISHED 2018.

Designed by Renata De Oliveira

Library of Congress Cataloging-in-Publication Data
has been applied for.

ISBN 978-0-06-268574-2

18 19 20 21 22 LSC 10 9 8 7 6 5 4 3 2 1

To four generations of Harvard freshmen in
"FRSEM 37u: Bob Dylan"
2004–2016

BOB DYLAN'S DISCOGRAPHY

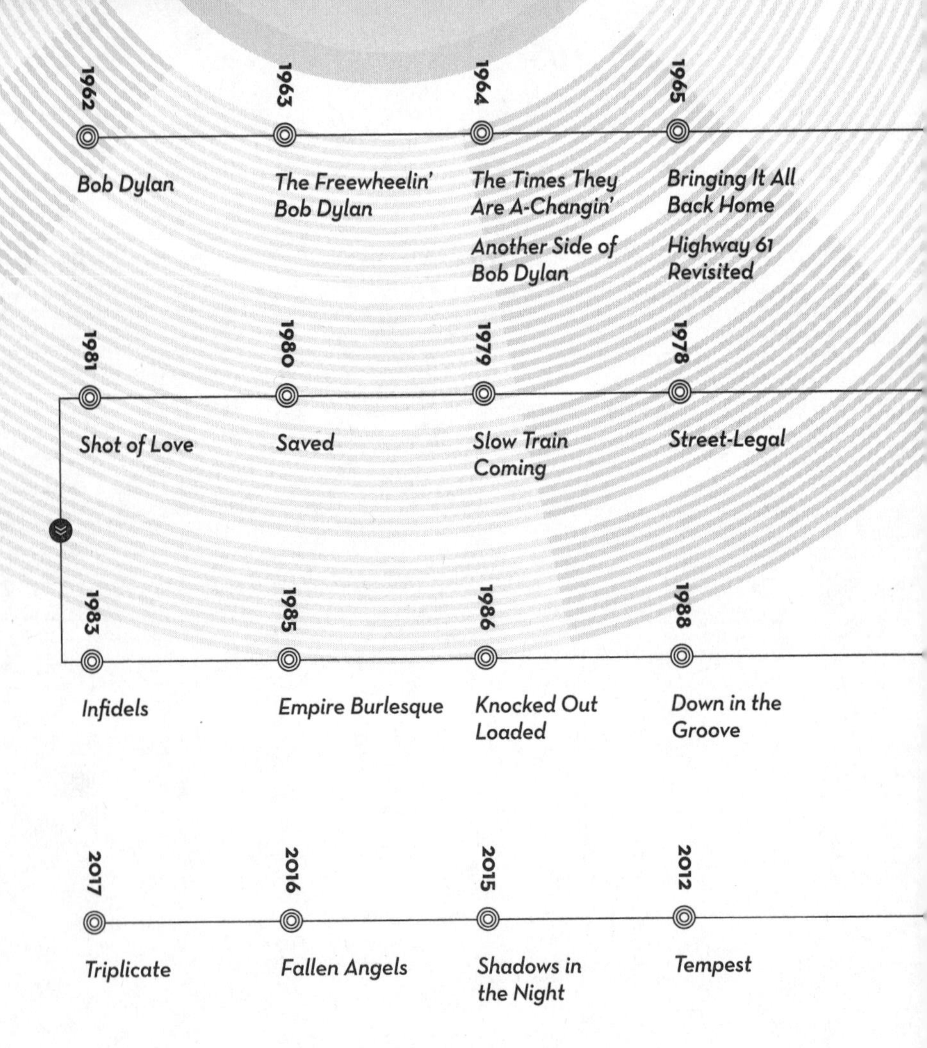

1966

Blonde on Blonde

1967

John Wesley Harding

1969

Nashville Skyline

1970

Self Portrait

New Morning

1973

Pat Garrett & Billy the Kid

Dylan

1974

Planet Waves

1975

Blood on the Tracks

The Basement Tapes

1976

Desire

1989

Oh Mercy

1990

Under the Red Sky

1992

Good as I Been to You

1993

World Gone Wrong

1997

Time Out of Mind

2001

"Love and Theft"

2006

Modern Times

2009

Together Through Life

Christmas in the Heart

CONTENTS

1

WHY DOES DYLAN MATTER TO US?

THERE'S A MOMENT WHEN ALL OLD THINGS
BECOME NEW AGAIN
—BOB DYLAN

Bob Dylan's *Blonde on Blonde* was one of two albums I packed in the trunk I sent from New Zealand to Ann Arbor, Michigan, in the early summer of 1974. The other was *Songs of Leonard Cohen*. I was twenty-three years old and had sold off the rest of my record collection to finance a two-month backpacking trip through Greece, before starting my doctoral studies at the University of Michigan. The trip to Greece was my first, but I had been fascinated by the Greeks and Romans since the age of nine, growing up in Auckland, New Zealand, half a world away from where their civilizations rose and fell. I arrived in Ann Arbor on August 18, 1974, days after Richard Nixon resigned from the presidency, ready to begin my professional life as a scholar and teacher of classical literature. My trunk finally arrived in October, and its familiar contents were a welcome sight. Along with the survivors of my record collection, that trunk contained the few classical texts I had accumulated as an undergraduate: the writings of Homer and Virgil, the epic poets of Greece and Rome, along with Sappho, Catullus, Horace, and Ovid, the brilliant lyric poets and love poets whose work captures what it means to live and love, to win and lose, to grieve and celebrate, and to grow old and die. For two thousand years, their poetry has fired the minds and imaginations of philosophers and poets, painters, sculptors and musicians, dreamers and lovers.

For the past forty years, as a classics professor, I have been living in the worlds of the Greek and Roman poets, reading them, writing about them, and teaching them to students in their original languages and in English translation. I have for even longer been living in the world of Bob Dylan's songs, and in my mind Dylan long ago joined the company of those ancient poets. He is part of that classical stream whose spring starts out in Greece and Rome and flows on down through the years, remaining relevant today, and incapable of being contained by time or place. That's how Dylan matters to me, and that's what this book is about.

From the beginning of his musical career, Bob Dylan has been working with artistic principles, and attitudes toward composition, revision, and performance, that bear many similarities to those of the ancients. He has also been living and writing in a world that bears many striking similarities to that of the ancient Romans, whose republic was the model on which the Founding Fathers built our own system. I believe that Dylan early on came to recognize this similarity, and it has been reflected in the worlds he creates for us in his music ever since.

Cullen Murphy's 2007 book, *Are We Rome?,* addresses this question, arguing that our time (Dylan's time) looks quite a bit like for the most part that of the Romans at various moments in their more than thousand-year history. According to Cullen, the ties that bind the two cultures include the condition of being a superpower, tensions caused by ethnic differences, the persistent memory of civil wars long after the last battle was fought, a sense of the fragility of political structures and decline of the human condition, the relaxation of moral and religious bonds, and a pushback against the countercultures.

At the heart of it, Rome around the last century BC and the beginnings of the first century AD and America in the second half of the twentieth share a sense of modernity, by which I mean a few things. By then Rome had more or less established herself as the dominant power in the Mediterranean world. The absence of serious external enemies, along with the sheer size of her empire, led to competing struggles among those whose task it was to govern the state and extend and defend its vast borders. Starting in the middle of the century these competing forces clashed, and a series of civil wars led to the elimination—by death on the battlefield, murder, and assassination—of one figure after another, along with the defeat of the ideals or interests they represented. The names are well known: Pompey the Great, Julius Caesar, Brutus, Mark Antony, all killed in a succession of bloody civil wars that went on for eighteen years. By around 30 BC, Augustus Caesar, as he would be renamed, the last man standing, delivered the final blow to the republic and stepped in as the first emperor of Rome.

This period of political uncertainty coincided with the emergence of a brilliant succession of poets and other writers, as happens at moments of political and national crisis or greatness: Athens in the fifth century BC, Elizabethan England, America and Great Britain between the two world wars—the rise of the so-called Moderns. Such moments give rise to a heightened sense of the past, along with uncertainties about the future. In each of these periods new art forms responded to what was happening, disrupting the old forms and traditions, busting them up, renewing what had gone before, moving into uncharted territory. The Roman poets in question will become familiar in the pages that follow: Catullus, Virgil, Horace, and Ovid. Others would have filled out their ranks, but their texts did not survive the centuries. Their art addressed the large issues of their day, the perilous state of their world, and the aftermath of civil war. Similarly, Dylan's art would speak to the horrors of the wars of his day, the Second World War and the cold war that followed, historic episodes like the Cuban missile crisis, and the fear of nuclear warfare, eventually Vietnam, even Iraq. And in both cases, through music and poetry that would prove to be enduring, memorable, and meaningful to ages beyond their own, Dylan and the ancients explore the essential question of what it means to be human.

Dylan's songs have been part of my song memory since my mid-teens, but it would be decades before they became more fully aligned in my mind with the Greek and Roman poets I was beginning to read back then. And it was chiefly in the twenty-first century that Dylan started to reference, borrow from, and "creatively reuse" their work in his own songs. I first began to make the connection after a trip to the coast of Normandy, where I had been invited in the spring of 2001 to give a lecture at the University of Caen on Virgil and other Roman poets. My host, Catharine Mason, a linguistics professor there, met me at the train station. She suggested that instead of touring the town, pretty much pummeled out of its historical state before the D-Day invasion of June 6, 1944, we might head for the beach. That sounded good to me, so I followed her to the parking lot. As we got into her car and she turned the key in the ignition, music came blasting from her car stereo. As we've all done, she had gotten out of the car earlier without thinking to turn down the volume, and the familiar bars of Dylan's "Idiot Wind," then and now one of my favorite songs, urgently interrupted our tentative conversation:

You hurt the ones that I love best and cover up the truth with
lies
One day you'll be in the ditch, flies buzzin' around your eyes
Blood on your saddle

Our conversation quickly turned to Dylan, to that song and its importance. For Catharine, a single mother who had recently gone through a divorce, and an American expatriate bringing up two young sons in France, the breakup song had powerful personal resonance. She had gotten hooked on Dylan twenty-five years after I had, with his 1990 album *Under the Red Sky,* whose nursery rhyme and fairy-tale traditions became part of the rhythm of bringing up her two young sons. As we walked on what had been Sword Beach, landing point for the British Third Division on D-Day, she talked about her plans for a conference on the performance art of Dylan. Did I want to give a paper at her conference, and maybe even co-edit the proceedings into a volume, she asked? I said sure, not really knowing how I would find a way into the topic. But in the back of my mind, I was thinking about how the songs from Dylan's 1997 album, *Time Out of Mind,* had lately begun somehow to remind me of the work of the Roman poets. Still, I had yet to share this insight with anyone.

It was not until many months after my trip to Caen, soon after September 11, 2001, the day that permanently changed the modern world, that what I would present at the Dylan conference became clear to me. Dylan's album *"Love and* came out on that day, and I bought it at the Tower Records in Cambridge, Massachusetts, in a daze, in the hours after the towers in New York had been leveled. When I eventually listened to the album, I heard Virgil, loud and clear in the tenth verse of "Lonesome Day Blues":

> *I'm gonna spare the defeated—I'm gonna speak to the crowd*
> *I'm gonna spare the defeated, boys, I'm gonna speak to the crowd*
> *I'm goin' to teach peace to the conquered*
> *I'm gonna tame the proud.*

The idiom, rhymes, and music of these lines belonged to Dylan, but the thought and diction, rearranged by Dylan, came from Rome's greatest poet, Virgil. In Dylan's lyrics, I recognized these lines from Virgil's *Aeneid,* spoken by Anchises, father of Aeneas, the mythical founder of Rome. Anchises, who had died on the journey from Troy to Sicily, instructs his son from the Underworld on just how Rome is to rule the world:

> but yours will be the rulership of nations,
> remember, Roman, these will be your arts:
> to teach the ways of peace to those you conquer,
> to spare defeated peoples, tame the proud.
>
> —Virgil, *Aeneid* 6.851–53, tr. Mandelbaum

Suddenly, when I heard Virgil's lines echoed in Dylan's song, my paper topic was obvious. "Bob Dylan's Performance Artistry" ended up taking place in Caen in March 2005, and I brought along my younger daughter, who was a college freshman at the time and a veteran of a few Dylan concerts. I was delighted, and surprised, that she'd chosen to spend her first college spring break with her father at a Bob Dylan conference. But that was the point. I'd taken her and her sister along to academic conferences before, where they would generally disappear into whatever city we happened to find ourselves. But at the Dylan conference in Normandy, things were different. My daughter attended every event, and even joined the Dylanologists for after-dinner Dylan trivia games. We were even treated to a Hendrix-style rendition of Dylan's "All Along the Watchtower" by Catharine Mason's teenage sons. I was astonished to see their enthusiasm, but perhaps I shouldn't have been. When I see the younger generation of Bob fans, like my daughters, or Catharine Mason's sons, or the students in my freshman seminar, engage with his work, it is a testament to the intergenerational nature of his work, and how his art endures.

In 2003, between my trips to Normandy, I decided to submit a proposal to the Freshman Seminar program at Harvard for a course on Dylan (the first of its kind, to my knowledge). The seminar was eventually approved by the faculty committee responsible for selecting these courses, though not without a fight. I later heard from a friend and member of the committee who had supported my proposal of pushback from some quarters. "What's he going to do, sit there and listen to 'Highway 61 Revisited' with his students?" was the general attitude. Well, yes, it would be hard not to include that song in the course. My friend had countered that my proposal was no different, and no less appropriate, than putting in to teach the works of T. S. Eliot. This argument won the day, and the seminar has been warmly supported ever since. I teach it every four years, most recently in the fall of 2016.

Since 2003, the seminar has evolved and changed, as Dylan has continued to produce new work and break new boundaries. We trace the evolution of Dylan's songs from their early folk, blues, and gospel roots and by way of the transition of his art from acoustic to electric in the studio and in performance, the latter being the arena that most inspires and motivates him. We move chronologically but also explore the way the themes of his song connect over time, are part of a larger system that connects song to song and album to album, down through the years. The themes comprehended by Dylan's songbook are as boundless as those of the folk and literary cultures from which his art emerged, and these are the themes of the seminar: music and social justice, war and the human response to war, love and death, faith and religion, song as compensation for the realities of mortality. I place particular emphasis on trying to have the students see Dylan's art as art and to attend to his songs not as autobiography, but as the product of a highly creative imagination that constantly manipulates and transforms linear time and the details of any actual life experience, much of which he has carefully concealed from the very beginning.

The first time I offered it in 2004, I had no idea what to expect. Would four or forty students apply for one of the twelve spots in the Dylan seminar? Would seventeen- and eighteen-year-olds actually be interested in the work of a musician who for many epitomizes only the 1960s, pretty far back in the rear-view mirror for them? I had over the years seen a certain number of teens and twenty-somethings at Dylan concerts, so I was hopeful. As it turned out, there is plenty of interest and every four years the crop has been plentiful. The students turn up for any number of reasons. Some want to understand the obsessions of a father or grandfather, others want to deepen their appreciation of Dylan's art or their own skills as songwriters. In the application, students are asked to say why they want to be in the seminar, and the responses show that Dylan's appeal is as varied as the dimensions of his art:

> *"I want to take this seminar because I want to be a better writer. I want to analyze his lyrics and internalize the reason people empathize with his sentiment. Maybe I can't, maybe it's innate."*
>
> *"I'm both a singer and composer I am interested in the way Dylan marries lyrics and music."*

> *"I knew all the words to 'Like a Rolling Stone' by the time I was four years old. My dad had played Dylan to me practically from birth."*
>
> *"My favorite English teacher from high school loves Dylan."*
>
> *"While Dylan may not possess the crazy guitar chops of guys like Slash or Jimi Hendrix, his lyrical genius makes his music just as, if not more, powerful."*
>
> *"I want to gain an understanding of how music can interact with history and philosophy."*

After twelve years of teaching the course, I have experienced firsthand the intergenerational power of Dylan. His art transcends time, and the power of his songs appeals to young adults whose parents were not yet born when Dylan started putting his words and music together. Dylan is here to stay. He has become a classic, each new album shifting the boundaries of the art he took up all these years ago. And with his 2016 award of the Nobel Prize in Literature, the world has recognized his literary merit—vindication for those who have long recognized the fact. My seminar happened to meet on October 13, 2016, the day the Nobel Prize was announced, and it was one of the high points of my teaching career to experience the utter joy of my first-year students on that day, since they knew the judgment of the committee was righteous.

This book will end with the Nobel ceremony in Stockholm, but for now let's visit a more recent stop there on Dylan's long path. On April 1, 2017, Bob Dylan, still on the road at the age of seventy-five, started a twenty-eight-concert European tour with the first of two performances at the Stockholm waterfront. Earlier in the day, he had met privately with twelve of the eighteen members of the Swedish Academy to receive his diploma and gold medal for the 2016 Nobel Prize in Literature, which had first been announced five months before, "for having created new poetic expressions within the great American song tradition." Sara Danius, academy member and its permanent secretary, reported the essentials of the private medal ceremony on her blog:

> Spirits were high. Champagne was had. Quite a bit of time was spent looking closely at the gold medal, in particular the beautifully crafted back, an image of a young man sitting under a laurel tree who listens to the Muse. Taken from Virgil's *Aeneid,* the inscription reads: *Inventas vitam iuvat excoluisse per artes,* loosely translated as "And they who bettered life on earth by their newly found mastery."

Danius's evasive "quite a bit of time was spent looking closely at the gold medal" fools nobody: everyone else in that room had seen the medal many times before. Dylan was clearly the one who must have been studying it most carefully. An artist who notices everything in the world around him, and one with a connection to Virgil's work, as we'll learn more about later in this book, Dylan would have been fascinated by the image on the reverse side of the medal, designed by Swedish engraver Erik Lindberg in 1902.

The man we see here is not just any young man. He would seem to represent the poet Virgil, one of the shepherd-singers of his poem *Eclogue* 1, "meditating the woodland Muse" as he sings in the shade of a tree. The singer on the medal is likewise looking up at the Muse as she plays the seven-stringed lyre, or *cithara* as the Greeks and Romans called it—the word that gives us *guitar*. Beside him is depicted an ancient box (*capsa*) with three papyrus rolls, the young man's supply of writing materials. Dylan knew just what he was looking at, having integrated

Homeric singing and lyre playing from the *Odyssey* into his 2012 song "Early Roman Kings"—"Take down my fiddle, tune up my strings"—which he would perform the next day in Stockholm. Like the image, the words engraved around the medal's rim are also Virgil's: *Inventas vitam iuvat excoluisse per artes*. In its larger context the line comes from a description of the privileged place that singers have deserved in Virgil's version of paradise in Book 6 of his epic the *Aeneid:*

> And Orpheus himself, the Thracian priest with his
> long robes,
> keeps their rhythm strong with his lyre's seven ringing
> strings,
> plucking now with his fingers, now with his ivory
> plectrum.
> And faithful poets whose songs were fit for Apollo
> **those who enriched our lives with the newfound**
> **arts they forged**
> and those we remember well for the good they did
> mankind.
>
> —Virgil, *Aeneid* 6, tr. Fagles

In 1945, T. S. Eliot wrote an essay titled "What Is a Classic?" devoted to Virgil, and to why the *Aeneid* became a classic over time. In 1948, when Eliot received his Nobel Prize in Literature, he must have been pleased to see Virgil's line of poetry, and the image, on the medal. What Eliot wrote of Virgil as classic in his essay could apply equally to his own work *The Waste Land*, the classic of modernist poetry, or to the work of Bob Dylan:

> [Virgil] was, if any poet ever was, acutely aware of what he was trying to do; the one thing he couldn't aim at, or know that he was doing, was to compose a classic: for it is only by hindsight, and in historical perspective, that a classic can be known as such.

This is a book about Bob Dylan, the genius of my lifetime in his artistic use of the English language, and of its song traditions—just as surely as Eliot was the poetic genius of the first half of the twentieth century. It is mildly ironic that Dylan has acquired this status. After all, the mention of Eliot in his 1965 song "Desolation Row"—a song he also sang on the Stockholm waterfront on the evening of the medal award—had an iconoclastic ring to it:

> *And Ezra Pound and T. S. Eliot*
> *Fighting in the captain's tower*
> *While calypso singers laugh at them*
> *And fishermen hold flowers*

In that song, Dylan may seem to be on the side of the calypso singers and fishermen, situated like them

Between the windows of the sea
Where lovely mermaids flow
And nobody has to think too much
About Desolation Row

As readers have noted, not only does Dylan name Eliot and Pound; in Eliot-like fashion his verse allusively builds on the ending of Eliot's first great poem *The Love Song of J. Alfred Prufrock* (124–31):

I have heard the mermaids singing, each to each.

I do not think that they will sing to me.

I have seen them riding seaward on the waves
Combing the white hair of the waves blown back
When the wind blows the water white and black.

We have lingered in the chambers of the sea
By sea-girls wreathed with seaweed red and brown
Till human voices wake us, and we drown.

Even though he seems in "Desolation Row" to distance himself from the two modernist poets, even as he alludes to one of them, like it or not, and like T. S. Eliot, Bob Dylan has also become an icon and a classic. Over that he has no control.

This is also a book about how Dylan's genius has long been informed by the worlds of ancient Greece and Rome, and why the classics of those days matter to him and should matter to all of us interested in the humanities. We live in a world and an age in which the humanities—the study of the best that the human mind has risen to in art, music, writing, and performance—are being asked to justify their existence, are losing funding, or are in danger of losing funding. At the same time, those arts seem more vital than ever in terms of what they can teach us about how to live meaningful lives. The art of Bob Dylan, no less than any other works produced by the human mind in its most creative manifestation, can be put to work in serving and preserving the humanities.

In his final treatise, *On Moral Duties,* written in 44 BC, Marcus Tullius Cicero, the Roman statesman, orator, writer, and thinker, quoting from the Roman playwright Terence, wrote: "I am a human. I consider nothing connected to humanity to be alien to me." For Cicero, thinking about justice and correct action in difficult times is a hallmark of humanistic thought, as is having empathy for the human condition. That was a mark of Cicero, and it is a mark of the focus on humanity that is at the core of Dylan's art. Dylan's art has long enriched the lives of those who listen to his music, through a genius that captures the essence of what it means to be human.

2

TOGETHER THROUGH LIFE

As was the case for many who came of age in the mid-1960s, my first Dylan experience centered on issues of social justice. When I was thirteen or fourteen, I sang "Blowin' in the Wind" with my school chorus in New Zealand. This song, from Dylan's first original album, *The Freewheelin' Bob Dylan,* was my first, true introduction to Dylan's music, though I was already familiar with the chart-topping version sung by the folk trio Peter, Paul and Mary. When we sang Dylan's version in chorus, I remember being somewhat put off by the "addition" of the words "Yes 'n' " before each of the questions, absent from the version I had hardwired from the radio—"Yes 'n' how many seas must a white dove sail?" and so forth. I recall pointing this out to the chorus director, who was clearly an early Dylan fan. He insisted that we sing the Dylan version, and I soon came to share his preference.

We could sing this song and make it our own back then in New Zealand because it belonged to no one time or place, but rather was right for any time or place as a cry for justice and peace. A few years later, in 1969, "Blowin' in the Wind" would take on a new and more immediate meaning for me, as I participated as a member of HART (Halt All Racist Tours), a student organization protesting the New Zealand rugby tours of apartheid South Africa. Now the song's reach had expanded, beyond the U.S. civil rights struggle that was its original backdrop. It was also about Nelson Mandela and others, sentenced by the apartheid regime to life imprisonment with hard labor on Robben Island—within clear sight of the pleasant beaches of Cape Town. Before 1967, New Zealand had been sending racially selected teams to play against whites-only rugby teams in South Africa. Then Maori players were allowed to play with the status "honorary white." "Yes, 'n' how many times can a man turn his head / Pretending he just doesn't see?"—those lines easily came to mind as I joined those marches. In these years the song also became an anthem as we demonstrated against my country's symbolically important and militarily insignificant involvement in Vietnam: "Yes, 'n' how many times must the cannonballs fly / Before they're forever banned?"

"Blowin' in the Wind" worked pretty well for those occasions, as it had for the seminal civil rights protest March on Washington on August 28, 1963, there sung not by Bob Dylan, but by Peter, Paul and Mary. Dylan himself sang two songs from his as-yet-unreleased album of that year, *The Times They Are A-Changin'*, first "When the Ship Comes In," accompanied by Joan Baez, and then on his own, "Only a Pawn in Their Game." In spite of its historical role at moments such as these, "Blowin' in the Wind" can't now, and couldn't then, simply be labeled a protest song. From his first known performance of the song, in April 1962 at Gerde's Folk City in Greenwich Village, Dylan struggled to free it from such categorization. Here's the way the songwriter—that's what he was, it needs to be stressed, not a protester—introduced the song back then:

> This here ain't a protest song or anything like that, 'cause I don't write protest songs. . . . I'm just writing it as something to be said, for somebody, by somebody.

Too late. As the Roman poet Horace (65–8 BC) put it, "Once it's let loose the word flies off and can't be called back." Dylan couldn't call back the song, but he soon stopped singing it in concert. He only performed it a handful of times during the decade in which it had become an anthem, including a memorable performance for his debut at the Newport Folk Festival on the evening of July 26, 1963. The memory of that show would help to fuel a sense of betrayal in the minds of folk and protest song purists not only at the next Newport festival in 1964, when Dylan opted not to play the song, but especially and finally at Newport in 1965, when he traded his traditional solo acoustic performance for one backed by Mike Bloomfield's electric guitar. Things had already changed at Newport the previous year, when Dylan played new, still-unreleased songs "It Ain't Me, Babe" and "Mr. Tambourine Man," the latter showing how astonishingly complex and poetic his language and song had become. The lyrics of "It Ain't Me, Babe" almost seemed to be directed to fans who had come expecting a repeat of "Blowin' in the Wind" from the year before: "Go 'way from my window / Leave at your own chosen speed / I'm not the one you want, babe / I'm not the one you need." The end of the refrain, "It ain't me you're looking for, babe," in hindsight, came across as a warning of what would happen at the next year's festival, on July 25, 1965. That evening, with Bloomfield's howling electric guitar riffs coming out of the darkness of the stage, alternating at times line by line with Dylan's singing, Dylan delivered the message: "Well I try my best / To be just like I am / But everybody wants you / To be just like them / They say sing while you slave and I just get bored / I ain't gonna work on Maggie's farm no more." The acoustic Dylan that folk fans loved hadn't gone away, but the song they had heard and loved two years before was silenced for the time being. Something else was happening that many folk purists, particularly older ones, along with those in the antiwar movement, were just not ready for. The songs of those years, culminating in the "thin wild mercury sound" of *Blonde on Blonde,* as Dylan himself described it, had entered another universe. In just fifteen months, from March 22, 1965, to May 16, 1966, Dylan recorded and released three albums—*Bringing It All Back Home*, *Highway 61 Revisited,* and *Blonde on Blonde*—that would establish and perfect the entirely new genre of folk-rock, a convenient label, even if it was not to Dylan's liking.

As the years went on, for people who stopped following Dylan's new music, who booed at the concerts of 1966 and were radicalized by the deepening involvement in Vietnam, Dylan was frozen in time, only those few songs relevant to what now mattered. He was useful as a protest singer, joined at the hip to that acoustic version of "Blowin' in the Wind." "Masters of War," the other great song on *The Freewheelin' Bob Dylan* that was adopted as a protest anthem, was addressed directly to those who build the weapons and send the young men to die for the wars from which they profit. Like "Blowin' in the Wind," it created an indelible association in the minds of those who would head off to Vietnam. Dylan's disclaimer about not writing protest songs made back in 1962 at Gerde's hadn't worked. His music, powerful from the beginning, would take on a life of its own. Dylan didn't turn up at the protest marches of 1965 and later to sing "Blowin' in the Wind," but singers like Joan Baez, Judy Collins, and others unknown and without fame would take over for him, at marches, in student unions, in student apartments, wherever the antiwar movement was to be found, in the United States, in Auckland, and throughout the world. As student protest leader Todd Gitlin put it: "Whether he liked it or not, Dylan *sang for us*. . . . We followed his career as if he was singing our songs."

Seven years after those songs came out, as Dylan relates in his memoir, *Chronicles: Volume One,* it was the message, not the music, that the media and general public recalled. By 1970 he had been off the road for four years, trying to dodge the demands of those who wanted him to return either to the antiwar song or to the sound and the lyrics that he had invented in 1964–66, the true basis of his musical fame by that point. After Dylan stopped touring in 1966, now raising his family with Sara Dylan in Woodstock and New York City, and writing very different music, as we'll see, he still couldn't get away from the old labels. In his memoir, Dylan recalls receiving his honorary degree from Princeton University in June 1970, annoyed at being labeled "the conscience of America":

> "Though he is known to millions, he shuns publicity and organizations preferring the solidarity of his family and isolation from the world, and though he is approaching the perilous age of thirty, he remains the authentic expression of the disturbed and concerned conscience of Young America." Oh Sweet Jesus! It was like a jolt. I shuddered and trembled but remained expressionless. The disturbed conscience of Young America! There it was again. I couldn't believe it! The speaker could have said many things, he could have emphasized a few things about my music.

For Dylan, it is the art of the song that matters. And song has powerful effects, especially when it responds to human conflict, to perceived injustice, to oppression. It is through song that we give depth to the sentiments for which mere speech is at times of crisis insufficient. And the more perfect the song, the more authentic the singer becomes in the minds of those who hear the song. How can a songwriter who creates songs with such fundamental and persuasive messages not believe those messages? That has always been the shackle from which Dylan has struggled to free himself. The message of a handful of Dylan's songs was what lingered in the consciousness of those who had heard them and had been involved, on one side or the other, for or against the war in Vietnam. To this day, the attitude of anyone old enough to have had a position on that war probably lines up pretty well with what they think of Bob Dylan the man, even after all these years.

"Blowin' in the Wind" is part of Dylan's poetry and art, exquisite in its classical structure and form. The song is written in three verses, each with three questions, each question extending over two lines, and each followed by the same answering couplet:

How many roads must a man walk down
Before you call him a ***man?***
Yes, 'n' how many seas must a white dove sail
Before she sleeps in the ***sand?***

Yes, 'n' how many times must the cannonballs fly
Before they're forever ***banned?***
The answer, my friend, is blowin' in the wind
The answer is blowin' in the wind

How many years can a mountain exist
Before it's washed to the ***sea?***
Yes, 'n' how many years can some people exist
Before they're allowed to be ***free?***
Yes, 'n' how many times can a man turn his head
Pretending that he just doesn't ***see?***
The answer, my friend, is blowin' in the wind
The answer is blowin' in the wind

How many times must a man look up
Before he can see the ***sky***
Yes, 'n' how many ears must one man have
Before he can hear people ***cry?***
Yes, 'n' how many deaths will it take till he knows
That too many people have ***died?***
The answer, my friend, is blowin' in the wind
The answer is blowin' in the wind

The urgent refrain that supposedly provides an answer really gives no answer at all, but rather creates its own questions: "The answer, my friend, is blowin' in the wind / The answer is blowin' in the wind." Is the wind blowing the answer away from us, never to be heard, or blowing it toward us, about to right the injustices of those nine urgent questions? Roman critics had a saying, "The art of poetry is to not say everything." That is precisely what Dylan's refrain does, indeed what much of Dylan's art does; it implants the possible answer in our imaginations, and the rest is up to us.

It is the height of irony that Dylan's main fear about how the song would be labeled may have lain elsewhere. One of his earliest and most famous interviews took place in May 1963, on Studs Terkel's radio show on WFMT in Chicago. The young Dylan was in town to play at a local club, the Bear. Not included in Terkel's 2005 published collection of interviews, *And They All Sang,* and therefore not to be found in Jonathan Cott's *The Essential Interviews,* the following exchange took place in the lead-up to Dylan's closing song and may be heard on a tape of the show:

ST: What's one way to sign off . . . a signing-off song?
BD: Sign-off song, let's see. Hmm. Oh, "Blowin' in the Wind," there's one I'll sing you.
ST: Isn't that a popular song, that's a popular song, I believe.
BD: God, I hope not.
ST: By "popular" I mean in a good sense. A lot of people are singing it.
BD: Oh, yeah.

After an eight-year hiatus, the song came back twice on August 1, 1971, in Dylan's afternoon and evening performances at the Concert for Bangladesh. "Blowin' in the Wind" returned to his setlists for good in January 1974. By then Watergate and Nixon were more in his fans' consciousness, and the song, obviously not a pop song, could finally take its place in his repertoire as part of his art in performance. I heard Dylan sing the song twice in 2016, in Boston in July and in Clearwater, Florida, in November and twice again in June 2017, on each occasion as the first of the two encores to close his concerts, its regular place in setlists of recent years—and he has now sung it at more than 1,400 concerts, a world away from the acoustic version recorded by the twenty-one-year-old Dylan. The song is still urgent in its questions, but it can't, couldn't ever, be attached to any one historical event or condition.

The same goes for "Masters of War," in terms of its enduring resonance. Dylan had performed the song 884 times by the end of 2016. Vietnam had become a dim backdrop by this time even in the minds of baby boomers, but the masters of war ("You that build the death planes / You that build the big bombs") never really go away. They were close at hand when Dylan performed acoustic versions of "Masters of War" in Australia and New Zealand in early 2003, including on March 15, when millions of demonstrators in those two countries and across the globe took to the streets urging two politicians not to proceed. American president George W. Bush and British prime minister Tony Blair had for months been building the case for bringing war to Iraq. I myself heard the song in those months reflecting on how relevant its message seemed, forty years on. After the bombs started falling on Baghdad on March 18, and in concerts for the rest of 2003, Dylan stopped playing the song. We'll never know why, but perhaps Dylan felt that would make it too overtly a "protest song," the old label. It returned the following year, however, and stayed on setlists until November 23, 2010, when it disappeared, so far for good, except for one performance on October 7, 2016, at the Desert Trip music festival in Indio, California, a weekend extravaganza where some of those particular fans would have expected to hear the song, along with hits by Neil Young, the Rolling Stones, Paul McCartney, Roger Waters, and the Who.

Throughout December 1974, as my first semester as a Ph.D. student was drawing to a close, I regularly stopped in at the local record store on campus to pick up Dylan's new album, *Blood on the Tracks,* unaware that Columbia Records had held up its release. My pilgrimages to the record store became part of the rhythm of life, and I made some friends in the process, leading to late nights throughout my Ann Arbor years with music and revolution in the air at a blues club called the Blind Pig, or the Del Rio, which offered free jazz on Sunday evenings. The Ann Arbor Blues Festival had debuted a few years before I got to campus, in the fall of 1969, and featured artists like Muddy Waters, B. B. King, Howlin' Wolf, Son House, and Lightnin'

Hopkins. There were funding issues and by 1974 the festival had finished its run, but there was still good music from local bands and musicians attracted to that town's entertainment market of more than thirty thousand students.

As fans would later discover, *Blood on the Tracks* was delayed because Dylan had gone back to Minnesota, where he re-recorded some of the songs. But in due course *Blood on the Tracks* turned up in January 1975 and soon took its place right up there with *Blonde on Blonde,* a new classic for a new decade. The characters of that earlier album had been mysterious and lovely: Louise and Johanna in "Visions of Johanna," the sad-eyed lady in "Sad-Eyed Lady of the Lowlands." The first girlfriend of my imagination had bits of each even before she materialized. After those eight years, things had changed with the romantic visions of *Blood on the Tracks*: "Situations have ended sad / Relationships have all been bad," Dylan sang on "You're Gonna Make Me Lonesome When You Go." He later denied that the album was about getting divorced from Sara Dylan. Sara Lownds had married Dylan on November 22, 1965, and their divorce would come almost three years after the songs were written. But there is no denying that with *Blood on the Tracks,* the art and the beauty seem to come more from a sense of hurt and loss, and seldom is experience not an ingredient of art, as Dylan himself has said. All these years later, the emotion in those songs is as palpable as ever, in the studio versions and thousands of versions recorded in concert. That is what literature, song, and the way they work on memory and experience conspire to give us. Poetry and music are compensations for the pain that comes along with the human condition, and they are what can help us along. That's what Virgil's words on the Nobel medal mean, honoring those "who enriched our lives with the newfound arts they forged."

The music that Dylan produced in the eight years between these two great albums indicates anything other than decline. But it's hard to articulate the disappointment back through those years that the particular sound of *Blonde on Blonde* had gone away, never to return. The music he made between that album and *Blood on the Tracks* was all part of Dylan's continuing evolution, particularly in mid-1967 as he worked with members of the Band, in seclusion in upstate New York. Some of this material was released on *The Basement Tapes* in 1975, and much of the rest was long available on unofficial bootleg versions, eventually to be released in 2014 in a six-CD set. Then came the relative simplicity of language on the 1967 album *John Wesley Harding,* with its biblical engagement and old-school feel. Here Dylan sang with a more spare accompaniment, turning away from the hip, mod sixties to a sound that seemed rooted in nineteenth-century Americana, a return to a new, creative version of the folk traditions that had always been in his blood. Eighteen months later, with the 1969 release of *Nashville Skyline,* Dylan seemed to be creating a new genre, now inventing country rock, as he had invented folk rock a few years earlier. The next year saw release of his album *Self Portrait*, and then

New Morning. *Self Portrait* was hit particularly hard by critics, including by music historian Greil Marcus, who famously opened his *Rolling Stone* review with the words "What is this shit?" It wasn't until 2013, when Dylan put out *The Bootleg Series Volume 10: Another Self Portrait,* with alternate, live, and overdub-free versions, that the brilliance of this period truly came to light, as Marcus himself would eventually acknowledge.

But the fact is that in 1975, when Dylan put out *Blood on the Tracks,* the world changed for those who cared about his music, maybe in part because of the sublimation of life experience into art, which is the essence of the album. Gone for now was the "old, weird America," as Marcus had so well described it, of the songs Dylan was laying down with the Band. Gone were the eighteenth- and nineteenth-century worlds of bootlegging, hoboing, and minstrel boys on *Self Portrait,* gone too the country pie of *Nashville Skyline.* And gone was the white picket fence that *New Morning* had tried to build around Bob and Sara Dylan and their four children, against the odds.

Many of the songs on *Blood on the Tracks* were constructed through the principles and practice of painting, a skill and insight he picked up from Norman Raeben, a painting teacher in New York City, in early 1974. To be sure, Dylan attributed to Raeben the very comeback that the album represented.

> I was convinced I wasn't going to do anything else,
> and I had the good fortune to meet a man in New

York City who taught me how to see. He put my
mind and my hand and my eye together in a way
that allowed me to do consciously what I unconsciously
felt . . . when I started doing it the first album I made
was *Blood on the Tracks*.

Dylan is characteristically vague on the actual methods or techniques, and one could claim that a song like "Visions of Johanna" from 1966 already seemed to reveal painterly qualities, but it is true that the vivid narrative technique in a song like "Simple Twist of Fate" from the new album gave it new effects that catch what he is talking about:

A saxophone someplace far off played
As she was walkin' by the arcade
As the light bust through a beat-up shade where he was
wakin' up
She dropped a coin into the cup of a blind man at the gate
And forgot about a simple twist of fate

In a radio interview with folksinger Mary Travers in April 1975, Dylan said of *Blood on the Tracks,* "A lot of people tell me they enjoy that album. It's hard for me to relate to that. I mean, it, you know, people enjoying the type of pain, you know?" That's the point, as Dylan, here deliberately disingenuous, well knew. His artistic genius—in his words, music, and voice—create pain, but precisely because of the brilliance of his art on this album, these songs produce recompense for the loss of love and the memory of what had once been. This is the quite intentional goal of songs like "Simple Twist of Fate," "Idiot Wind," or "If You See Her, Say Hello." These songs also hold the trace of a hope that all might not be lost: in "Simple Twist of Fate" the man "Hunts her down by the waterfront docks where the sailors all come in / Maybe she'll pick him out again," this also giving the point of view of the character in the song; or the switch at the end of "Idiot Wind" from "You're an idiot, babe" to "We're idiots, babe." Sharing the blame; or at the end of "If You See Her, Say Hello," "Tell her she can look me up, if she's got the time"—though in other versions, any hope is pretty remote, as we'll see. To have lived through more than forty years with all of the music and poetry of these songs, from the album and in performance, is a source of good fortune and of genuine pleasure and deep contentment, even—or especially—with the pain the album so exquisitely expresses.

Much of the album focuses on nighttime, the time of day when the relationships in its songs seem to fall apart, perhaps also the case with Dylan's real-life relationships. The first line of the first song of the album, "Tangled Up in Blue," seems to start on a bright note: "Early one mornin' the sun was shinin' / I was layin' in bed," but within a moment that feeling is illusory, as the relationship is suddenly no more: "Wonderin' if she'd changed at all / if her hair was still red." The singer's early-morning memory eventually gets back to the evening breakup, after driving out west in a car that the couple abandons as they "Split up on a dark sad night / Both agreeing it was best." The next song, "Simple Twist of Fate," begins with a twilight encounter, now in third-person narration: "They sat together in the park / As the evening sky grew dark." After a one-night stand that could in the narrator's mind have led to something, in the morning he finds that she's gone: "He woke up, the room was bare / He didn't see her anywhere." In "Meet Me in the Morning," morning and night frame the song, which begins with "Meet me in the morning, 56th and Wabasha," and ends with the "sun sinkin' like a ship," and in between "They say the darkest hour is right before the dawn." The year after Bob and Sara Dylan's divorce, finalized on June 29, 1977, Dylan seemed to recall this aspect of the album's songs: "I don't have anything but darkness to lose. I'm way beyond that." A good deal of the melancholic and painful power of this album, whatever the realities of Dylan's personal situation, comes from these moments, all shadows in the night, a time of day that would continue to be the temporal setting and condition for the best of Dylan's song.

In the beautiful "If You See Her, Say Hello," the narrator's memories of what has been lost in the relationship also come as night falls, in the second verse: "But to think of how she left that night, it still brings me a chill," and then again in the final verse: "Sundown, yellow moon, I replay the past." I single out this song for its intense lyric qualities, and not so much for the painterly qualities so apparent elsewhere on the album. The gift of lyric poetry resides in its ability to precisely capture the condition of individuals in their sorrows, joys, loves and losses, desires, hatreds and jealousies. Such poetry is intimately connected to song—again, *lyric* from *lyre,* the guitar of the Greeks and Romans. Like song, it enables us to read ourselves into the situations that the poetic voice creates in aesthetically compelling modes. "If You See Her, Say Hello" is another such lyric song-poem. Its five verses are an elaboration of the title, a request by the singer for someone to say hello to a woman who walked out on him, its first verse closing with the rawness of the singer's feelings: "She might think that I've forgotten her, don't tell it isn't so."

This song actually exists in two versions, and in performance with many variations. The first version that Dylan released, on *Blood on the Tracks,* as quoted above, was actually recorded on December 30, 1974, effectively revising and for many years canceling out an earlier version, which was recorded in September 1974 and eventually released in 1991 on *The Bootleg Series Volumes 1–3*. In this version, the messenger was a rival, at least in the singer's imagination: "If you're makin' love to her, kiss her for the kid." The change makes a world of difference, as the element of jealousy complicates things and makes it a different song, as do multiple other changes in yet other versions, including the first known live performance, at Lakeland, Florida, on April 18, 1976, during the second Rolling Thunder Revue, as the Dylans' marriage was falling apart: "If you're making love to her, watch it from the rear / You'll never know when I'll be back, or liable to appear." Other parts of the song change in different performances over the years, replaced by brilliant absurdist lines, "Her eyes were blue, her hair was too, her voice was sort of soft," or a brutal closing to the song in concert in 2002: "If she's passing back this way, and it couldn't be too quick / Please don't mention her name to me, you mention her name it just make me sick." And the third stanza is gone altogether, absent from official collections of Dylan's lyrics. But what I heard all those years ago in Ann Arbor was the pure, lyrical version from the 1975 album, and that's the one that has stayed with me over the years: "Tell her she can look me up if she's got the time."

LEAVING AND COMING BACK TO DYLAN

In the fall of 1977, I left Ann Arbor for Cambridge, Massachusetts, to begin teaching in the green pastures of Harvard University, about fifteen years after Dylan met folksinger Eric Von Schmidt in those same pastures. By then my Dylan collection had been topped up. Dylan's Muse also seemed to have returned in what looked somewhat like a second classic phase, matching the first from a decade earlier. In early 1976, he had released *Desire*—not quite up to *Blood on the Tracks,* but fine enough. With Dylan's next album, *Street Legal,* appearing in 1978, the winds of change were again beginning to shift in his music. The opening lines of its first song, "Changing of the Guards," are vivid and allusive: "Sixteen years / Sixteen banners united over the fields," inviting the listener to look back those sixteen years to the beginning of Dylan's career, and take stock of how far he'd come and think in the apocalyptic lyrics of the song about where he might be headed: "But Eden is burning, either brace yourself for elimination / Or else your hearts must have the courage for the changing of the guards." There are some good songs on this album, chiefly for me "Is Your Love in Vain," but the album as a whole was flawed, as Dylan clearly felt by the best index available: only one song, "Señor," truly entered the repertoire of Dylan performances, and most he didn't play after 1978.

For the most devoted Dylan fans who have followed his music through each new stage, his songs and all they evoke become a part of us, with each new album adding another layer. For other fans, he effectively disappeared at various points, starting in 1964 or 1965, quitting their world of folk and protest songs to create a different kind of art. To these fans, Dylan had sold out to a hipster look, and had traded acoustic for electric, with all that connoted for the causes with which they had identified him. But what he gave them in those first two years endured, along with the bittersweet memory of what he had been to them, kept alive by new covers of those particular songs by generations of folksingers who came after. Some disappointed fans stuck around through 1966, hoping that Dylan's sound, which alternated in performances of that year between solo acoustic and electric with supporting musicians, would return to the former. When it didn't, these people booed at his concerts, and eventually either came to see what was happening there and found something in it that made sense, or decided to leave for good.

The next crop of Dylan fans to take their leave did so for different reasons, in 1979, when the changing of the guards had come to pass as he started writing and singing Christian songs, often preaching from the stage about hellfire and damnation before launching into his performance. That version of Dylan just didn't fit in with where they were in their lives or what they believed, or didn't believe in, or with the Dylan they thought they knew from 1966 or 1975 or some other moment. And so it has continued with Dylan's constant evolution through the decades, with some fans disembarking and others coming back on board, and newer, younger ones signing up for the first time. It is an essential part of Dylan's genius that he is constantly evolving as an artist. This is not true of the artists of similar longevity, say Leonard Cohen, Joan Baez, Joni Mitchell, Neil Young, Van Morrison, or Bruce Springsteen. Inevitably that constant evolving creates periods of experimentation and exploration, some less successful than others, but always moving restlessly toward something, and with the music of the last twenty years now having reached, and sustained, a third classic period.

Dylan's art works in elemental ways, not just through his words and music and voice, but also through his look and appearance. This is also part of his art, from his look of youthful, potent frailty in his early twenties, to his hip and sexualized look on the 1966 tour, through to the powerful maturity of his middle years. His look during the Rolling Thunder Revue tours of 1975–76, which you can see on YouTube and in the 1976 TV movie *Hard Rain,* is part of the appeal of those performances: 1970s hipster in his mid-thirties, dressed in denim and leather, sometimes sporting a bandana or turban, sometimes with an ornate floral arrangement in his hatband, frequently with white face paint, or with a straggly beard. And into recent years with his elegant, expressive, weather-beaten face, and his scrupulous attention to costume: outfits and hats that at times turn him into a Civil War officer, at times a cowboy, at times the vaudeville performer. In all of these evolutions there is an enigmatic presence that can't quite be comprehended or described. With Dylan, everything is performance, and all aspects of performance—the words, the music, the voice, the bands, and the look—coming together to create the unique phenomenon that is Bob Dylan.

3

DYLAN AND ANCIENT ROME: "THAT'S WHERE I WAS BORN"

GOIN' BACK TO ROME / THAT'S WHERE I WAS BORN.

—BOB DYLAN, "GOING BACK TO ROME," 1963

IF YOU WERE BORN AROUND THIS TIME OR WERE LIVING AND ALIVE, YOU COULD FEEL THE OLD WORLD GO AND THE NEW ONE BEGINNING. IT WAS LIKE PUTTING THE CLOCK BACK TO WHEN BC BECAME AD.

—BOB DYLAN, *CHRONICLES: VOLUME ONE*, P. 28

In March 2007, I traveled to the University of Minnesota for a symposium in Bob Dylan's home state titled *Highway 61 Revisited: Dylan's Road from Minnesota to the World*. The conference was designed to coincide with the exhibition *Bob Dylan's American Journey, 1956–66,* concurrently taking place at the university's Weisman Art Museum. Many of the best-known Dylan scholars were in attendance: Michael Gray, C. P. Lee,

Greil Marcus, Christopher Ricks, Stephen Scobie. The symposium was evidence that Dylan had become part of the academic mainstream. But that fact alone was not what drew me to the north woods along with the other Dylanologists. It was something more: the opportunity to come together to discuss Dylan in this place where his genius had first come into being, at the university where Bob Zimmerman was technically enrolled in 1959–60, just a few blocks from Dinkytown and the coffeehouses where he began in earnest to practice and perfect the art he would take out into the world.

The day before the conference, like many of the others attending, I signed up for a guided bus tour of Hibbing, Minnesota, the town where Dylan grew up. Hibbing is situated about seventy miles northwest of the city of Duluth, built on the rich iron ore of the Mesabi Iron Range, and at the edge of the town lies the world's largest open-pit iron mine. Dylan was born in Duluth on May 24, 1941, and grew up in Hibbing after his family moved there when he was six years old. The bus ride itself was memorable and scenic, as we headed north from Minneapolis on Highway 61, the road that follows the Mississippi all the way down to New Orleans, and rode through pine stands, past the Frank Lloyd Wright gas station in Cloquet, then on into Hibbing. We were a busload of about forty-five Dylanologists and assorted Dylan fans, including a young guy whose name tag read Jack Fate—the character played by Dylan in the underappreciated 2003 film *Masked and Anonymous,* as he was eager to explain to the few who needed explaining. Jack was handing out Highway 61 bumper stickers.

We eventually found ourselves standing in the library of Hibbing High—the magnificent "Castle in the Wilderness," as it's known—from which Robert Zimmerman graduated in 1959. Our tour guide, John "Dan" Bergan, a now-retired English teacher at Hibbing High, had been a classmate of Dylan's younger brother, David Zimmerman. David graduated five years after Bob and was "a terrifically talented musician in his own right," according to Bergan. Our busload of pilgrims was also treated to a talk by eighty-three-year-old B. J. Rolfzen, who had once been Dylan's English teacher. You could tell he must have been a dynamic teacher fifty years earlier, engaged by poetry and with a fire for conveying the magic of literature to his students. Music journalist and cultural critic Greil Marcus has described this moment from Rolfzen's talk:

> Presumably we were there to hear his reminiscences about the former Bob Zimmerman—or, as Rolfzen called him, and never anything else, Robert. Rolfzen held up a slate where he'd chalked lines from "Floater," from Dylan's 2001 *"Love and Theft":* "Gotta sit up near the teacher / If you want to learn anything." Rolfzen pointed to the tour member who was sitting in the seat directly in front of the desk. "I always stood in front of the desk, never behind it," he said. "And that's

> where Robert always sat." He talked about Dylan's "Not Dark Yet," from his 1997 *Time Out of Mind:* "I was born here and I'll die here / Against my will." "I'm with him. I'll stay right here. I don't care what's on the other side," Rolfzen said, a teacher thrilled to be learning from a student. With that out of the way, he proceeded to teach a class in poetry.

The Hibbing experience was all part of what later came to seem to me a carefully staged tour. It reminded me of a visit I'd taken a few years earlier to Max Gate, the house that novelist and poet Thomas Hardy designed and lived in on the outskirts of Dorchester in Dorset, England, from 1885 till his death in 1928. Or else it was a bit like visiting the Mark Twain House in Hartford, Connecticut. As a 2016 headline in the *CTPost* put it, "Mark Twain fan visits his Hartford mansion, finds it's like communing with a long-lost friend." Whatever we think we are doing on such journeys, what moves us is the sense of being at the wellspring of artistic creation, where creative genius began to form the art that would become central to our own lives and imaginations. In Hartford, we're looking for Huck or Tom. In Dorsetshire, we're hoping to run into some sign of Tess or the mayor of Casterbridge. Likewise, in Hibbing, we were all there looking for something to connect us to the Dylan we had known back in our youth and been with ever since. We were hoping to find it in the magnificent Hibbing High auditorium, where the fifteen-year-old Bob Zimmermann had played with his band, singing and pounding out a Little Richard tune on the piano, as recalled by his then friend John Bucklen:

> He got up there . . . in this talent program at school, came out on stage with some bass player and drummer, I can't remember who they were, and he started singing in his Little Richard style, screaming, pounding the piano, and my first impression was that of embarrassment, because the little community of Hibbing, Minnesota, way up there, was unaccustomed to such a performance.

I think we could all imagine that event, but in 2007, fifty years after the show, it was hard to get close. Bob wasn't there, but it was also easy to imagine him up on the stage looking out at the audience in the elegantly upholstered seats of the 1,805-capacity auditorium of which Dan Bergan, who wrote a booklet on the school, rightly noted, in language that, like the auditorium, seemed remote from the hard realities of the Iron Range:

> Nowhere in the United States can one find a high school auditorium—perhaps any auditorium—of such incomparable beauty, of such ornate and elaborate decoration . . . the auditorium features a 40- by 60-

> foot stage, framed by its 20- by 40-foot proscenium arch whose borders are marked by massive pillars with composite capitals in gold rising on each side of the stage.

Dylan would soon enough be performing at Carnegie Hall in New York and at the London Palladium, but that stage in Hibbing was not a bad place to start. This auditorium must be emblazoned in his mind. The nostalgia involved in the activation and exploration of memory is something that is essential to Dylan—as he said in 1967, "You can change your name / but you can't run away from yourself."

After visiting Hibbing High, our group, a little ragged from the warmth of the early spring day, made the short three-block walk from the school down Seventh Avenue, now "Bob Dylan Drive," to the corner of Twenty-Fifth Street, and the house Bob Dylan grew up in. According to the Iron Range Tourism Bureau, it is no longer open to the public—"drive-by visits only"—but on that day the owner had actually opened its doors and allowed us to go into the front living room, where he had set up a display of Dylan memorabilia on a coffee table. There was a Dylan song playing, I can't quite remember which one, and I think all of us felt a combination of pleasure at having arrived at such a place, along with slight embarrassment to be intruding in the inner sanctum. I was relieved that a request to visit the bedroom was declined, though some went around the side of the house to look up at its window. The owner of the house told us about Dylan's own occasional visits over the years. He would spend time up in the bedroom of his old house, presumably making contact with memories of listening on the radio to the music that would form him, first gospel blues and country, later rock and roll. He surely found his teenage self on these occasions.

Lunch was at Zimmy's, which has since closed as the town continues its economic decline. Some of us bought very unauthorized-looking Zimmy's T-shirts, along with copies of B. J. Rolfzen's memoir, *The Spring of My Life,* a self-published book in ninety-five pages of Courier font—and an interesting account in its own right of growing up poor in post-Depression America. The bus also took us a few miles out of town for a visit to the famous iron ore pit that you can see from the moon. The best ore was long gone, even when Dylan was growing up, and it was easy to connect to the song "North Country Blues" from *The Times They Are A-Changin'*—a mining blues folk song Dylan would sing at the Newport Folk Festival on July 26, 1963, then once again, for the last time at a concert, at Carnegie Hall, on October 26 of the same year. "This is a song about iron ore mines, and—a, iron ore town," he said at Newport. The song is in the voice of a woman, as we discover only in the fourth verse, brought up by her brother, who falls victim to the mines, following the same end as her father. In a final blow her husband deserts her and her three children. Dylan had written the song following a trip back to Hibbing, before the public discovered that he had grown up in the town. Andrea Svedberg broke the news of that reality in a *Newsweek* article published the Monday after the Carnegie Hall concert.

Once the Hibbing connection was made, "North Country Blues" was too easily situated in Hibbing and to the background of Bob Zimmerman, despite its narrator's female voice and the far different details of its story. Maybe that was why Dylan sang it only once more, in 1974 at a benefit concert for the Friends of Chile. By 2001, when "Floater (Too Much to Ask)" came out, Dylan cared less about people knowing where he came from, and B. J. Rolfzen in his talk is not the only one to have detected autobiographical undertones to the song, both in the lines he quoted and in the ending of the same verse, on the young people of the town:

They all got out of here any way they could
The cold rain can give you the shivers
They went down the Ohio, the Cumberland, the Tennessee
All the rest of them rebel rivers

By the time of that song, 2001, Dylan's real identity and background was even more beside the point. While "North Country Blues" is a song that can be tied to the hard lives of those who worked and died in the mines of Hibbing, Minnesota, it is even more a song that came more from the folk tradition of mining songs, and especially from the fertile mind of Bob Dylan. Like Dylan, our group soon enough boarded the bus and headed south, following his fifty-year-old trail, to the University of Minnesota, and the next day for coffee in Dinkytown, where he went in the fall of 1959 to take up the art of folksinger performance on his way to Greenwich Village and destiny. The conference itself was memorable enough, but what has stuck in my mind most is that day, spent in the little Minnesota town of Hibbing.

LATIN AND THE LATIN CLUB, HIBBING, 1956–57

As the only classicist in the group, I was also in Hibbing looking for something else, for traces of a bond I shared with Bob Dylan that for me dated back to 1959, when I began studying Latin at the age of nine. Following lunch at Zimmy's, I slipped out and walked the two blocks to the Hibbing Public Library. One of the waitresses had told me there was a Dylan exhibit there, featuring a copy of the *Hematite,* Dylan's high school yearbook from 1959, the year he graduated. The *Hematite* was named for the mineral form of iron oxide that brought wealth to the town, and had in the days before the main lode dried up paid for the building of its magnificent school. I had already seen page 76 of the yearbook, at a Dylan exhibit in Seattle in 2005, and in the Scorsese documentary *No Direction Home,* so I knew what to expect. On that yearbook page the life and career of the future Nobel laureate was summed up in just three details:

Robert Zimmerman: to join "Little Richard"–
Latin Club 2; Social Studies Club 4.

Plenty has been written about Bob's early interest in Little Richard, one of the foundational singers of rock and roll, whose hit "Tutti Frutti" shot up in the charts at the end of 1955, when Bob was a freshman at Hibbing High. By the following fall, backed by the Shadow Blasters, his name for the first band he had put together, Bob Zimmerman was himself now imitating the songs and stage antics of Little Richard. Indeed, the head shot of Bob Zimmerman at the top of that same yearbook page even alluded to the identity his notice craved, in the form of his trademark Little Richard pompadour hair style. This was well before he started taking on the persona, and the look, of Woody Guthrie as he headed for the folksinging scenes of Greenwich Village.

But few have paid much attention to his membership in the Latin Club. With his newfound performing interests, and from the evidence of his dropping off the honor roll from 1956 to 1958—he made it back on for his last year—his later claim to be interested in nothing beyond his music (liner notes, *Biograph,* 1985) might seem credible enough, though mostly on piano, not yet guitar. But right around this time he was also turning up to Latin class and to Latin Club meetings, and he certainly posed for the group photo of the club that came out in the 1957 *Hematite*. Bob Zimmerman's enrollment file "disappeared" years ago from the meticulously kept records of the school, but we know that he was taking Latin and learning about Rome that same year he put his first band together. In addition to the yearbook, the school paper, the *Hibbing Hi Times,* for November 30, 1956, in the regular "Club Notes" column also gives us a unique rarity, a record, unimpaired by the potentially creative memory of those friends who later recalled this or that detail—part of a day in the life of the fifteen-year-old:

> **SOCIETAS LATINA HOLDS INITIATION**
>
> Societas Latina [Latin Club] held its annual initiation party and ceremony for new members recently in the high school cafeteria. Several associated members of the club were present also.
>
> Second-year students vied on a mock TV program, answering questions on Roman gods and goddesses and identifying words dealing with various phases of Roman life. Winners were awarded prizes. After the formal pledge of allegiance by new members, initiates received badges and were raised from the status of slave to that of plebeians. Members then adjourned to the punch bowl where Consul Mary Ann Peterson and Anna Marie Forsmann, in Roman dress, presided.
>
> Consul Joe Perpich, assisted by Dennis Wickman, **Bob Zimmerman**, and John Milinovich, was in charge of the formal induction and radio program.

For whatever reason, interest in the Roman gods and goddesses, helping with the radio, or the favorable gender imbalance (fifty girls to fourteen boys)—or all three—Bob Zimmerman was a member of the Hibbing High Latin Club. The only other information about the Latin Club comes with the paper's issue for March 15, 1957, in the spring of Bob's membership year, under the headline LATIN CLUB EDITS IDES OF MARCH NEWS:

> Societas Latina members today published a paper to celebrate the death of Caesar on the Ides of March (March 15). The paper included Roman history, an original poem, cartoons, and many other items with a Roman background.

Any trace of that paper is long gone, but it is safe to assume Bob Zimmerman played some role in the celebration. Almost sixty years later, as we'll see, Dylan was quoted as saying, "If I had to do it all over again, I'd be a schoolteacher—probably teach Roman history or theology."

We can't be sure what got Bob Zimmerman interested in Latin and the Romans, but it looks as if those interests started in the years before he walked into Miss Irene Walker's Latin class in the fall of 1956. Bob's uncle owned the Lybba, named after Dylan's great-grandmother, one of the town's four movie theaters, along with the State, and the Gopher, like the Lybba both just a few blocks from his home, the fourth a drive-in. The early to mid-1950s saw an intensification of movies about Greece and Rome, the latter in particular, along with biblical movies, with or without Romans. This was part of a post–World War II, Cold War–generated escape into the relative security of antiquity: swords and sandals, rather than the atom bomb. At the same time, these years saw the height of McCarthyism and the blacklisting of Hollywood actors, producers, and directors. The ancient world could be used as a medium for camouflaging contemporary red-baiting while depicting persecutions emanating not from Washington, D.C., and the House Un-American Activities Committee, but rather from the city of Rome: between 1950 and 1956, when Bob decided to take up Latin, any number of such movies were available for him to have seen, including Joseph L. Mankiewicz's 1953 hit version of Shakespeare's *Julius Caesar,* starring Marlon Brando, one of Dylan's favorites, who got the best actor nomination for his role as Mark Antony.

In these years the following movies about the ancient world were available for Bob Zimmerman to see, free at the Lybba, or at either of the other two theaters, opening on the following dates:

Serpent of the Nile: Gopher, July 26, 1953
The Robe: State, January 1, 1954 (and its sequel):
Demetrius and the Gladiators: Lybba, June 24, 1954
Julius Caesar: State, February 9, 1955
The Silver Chalice: State, February 11, 1955
Jupiter's Darling: Lybba, March 11, 1955
Helen of Troy: State, March 4, 1956
Alexander the Great: Lybba, June 16, 1956

In 1951 he may have been too young for *Quo Vadis,* with Peter Ustinov as the lyre-playing emperor Nero, but it probably made a return visit in the years that followed. By the time *Ben-Hur* came out in 1959, Bob Zimmerman was moving on, though he claimed in an interview that the book on which the movie was based was part of the scriptural reading he did in his youth, just as he mentions *The Robe* and the 1961 *King of Kings* as early influences. There is not much else to do in Hibbing, particularly in the cold of the northern Minnesota winter, whether or not the theater is owned by your uncle.

I know I'm not the only classicist who was attracted to the world of Rome by Stanley Kubrick's 1960 movie, *Spartacus,* starring Kirk Douglas, which I first saw as an eleven-year-old. That movie opened at the Lybba on December 29, 1961, when Bob Dylan was back in Hibbing from his first year in Greenwich Village, for the end-of-year holidays—a year later he chose to visit Rome, and on his return to Greenwich Village sang a song he had just written, "Goin' Back to Rome." These movies were beginning to peter out when Elizabeth Taylor and Richard Burton gave us Cleopatra and Mark Antony in Mankiewicz's lavish 1963 epic, *Cleopatra*. Such things happen. Bob Zimmerman moved on, dropped Latin and stuck with his music, and became Bob Dylan. But my contention is that the memory of his contact with classical antiquity, like the memory of everything else, stayed with him, and had a similar early influence on the evolution of his music, as did the poetry he read in B. J. Rolfzen's English class and his own extensive and varied reading.

According to Dylan's own account in *Chronicles: Volume One,* published in 2004, the Rome of Hibbing makes one more appearance in his high school days, by way of the *Black Hills Passion Play of South Dakota,* a touring group that came to town to act out the suffering, crucifixion, and resurrection of Jesus. It seems they also needed locals to play the part of extras, as Dylan fondly recalls:

> One year I played a Roman soldier with a spear and helmet—breastplate, the works—a non-speaking role, but it didn't matter. I felt like a star. I liked the costume. It felt like a nerve tonic . . . as a Roman soldier I felt like a part of everything, in the center of the planet, invincible. That seemed a million years ago now, a million private struggles and difficulties ago.

Who knows what year this was, perhaps Dylan's sophomore year of high school, when members of the Hibbing High Latin Club got to take on such roles. If he was a Roman soldier, he presumably participated in the scene depicted in the gospels where Roman soldiers cast lots to see who will get the tunic of the crucified Jesus—both scenes familiar to him from *The Robe* and *King of Kings*. Bob Dylan revisited that scene in the 1975 song "Shelter from the Storm," where the singer's role is different, but reminiscent of the play he refers to in *Chronicles.*

First "she walked up to me so gracefully and took my crown of thorns," suggesting an identification with Jesus Christ, and four verses later "In a little hilltop village, they gambled for my clothes / I bargained for salvation an' they gave me a lethal dose." It doesn't matter whether his role as a Roman soldier was a reality or one of the many inventions and embellishments in his memoir, though the former seems more likely in this case. In his mind, back in 1957 and an epoch later in 2004, the road from Hibbing, like all roads, led to Rome. Dylan went back to Rome again, and to his role as a Roman soldier, in his Nobel lecture, delivered on June 5, 2017. In the lecture, he discusses three books that influenced him since grammar school, *All Quiet on the Western Front, Moby-Dick,* and the *Odyssey,* and describes the experience of Paul Bäumer, the soldier-narrator of *All Quiet* as being like "You're on the real iron cross, and a Roman soldier's putting a sponge of vinegar to your lips."

In Dylan's 2006 song "Ain't Talkin'," the narrator says, "I'll avenge my father's death, then I'll step back." While the avenging of a father's death may initially suggest *Hamlet,* one of Dylan's favorite plays, I believe the echoes of the line may also lead to Rome, and to the aftermath of the killing of Julius Caesar on the Ides of March, 44 BC, the event celebrated by the Latin Club in 1957. As is now well known, "Ain't Talkin'" steals a number of verses from the exile poems of the Roman poet Ovid, banished in AD 8 by the emperor Augustus to the desolate shores of the Black Sea. When Augustus took control through civil war and came to rule over the Roman Empire, he presented himself as restoring the state from the slavery imposed by Brutus and the other assassins of Julius Caesar:

> Those who killed my father I drove into exile, by way of the courts, exacting vengeance for their crime. . . . I did not accept absolute power that was offered to me.

The reality was otherwise, of course. Augustus maintained the trappings of republic, but in effect his power was absolute; he avenged his father's death, but he did not step back.

Whatever the impulse, for Bob Dylan the city of Rome, and along with it the culture of the ancient Romans, came to hold a special place over the years. We'll never know for sure what all those movies and his membership in the Latin Club have to do with this productive association, but the fact is that Rome and the Romans turned up in his songs from early on, and they continue to play a role in his creative imagination.

DYLAN AND CATULLUS

Folk music and the blues may be seen as the primary reservoir of Dylan's words and melodies for pretty much all of his music that followed. Rock and roll was the musical staple of his high school years, and it remained a part of him as he soaked up the various folk traditions, in Dinkytown in Minneapolis, and later in Greenwich Village. But folk was the old from which the new would emerge. For the youth of America, rock and roll was generational; it belonged to them. It cleared out the music of their parents, the era before immediately after World War II, the Great American Songbook, given voice by Perry Como, Frank Sinatra, and Tony Bennett—the mine to which Dylan would return, starting with the 2015 album *Shadows in the Night*. With what was happening, musically and culturally in the mid-1960s, Bob Dylan's genius was in the right time and the right place.

Something similar was happening in the middle part of the first century BC in Rome. Traditional forms of literature, drama, and early epic poetry were coming to be perceived as old-fashioned, precisely as society was opening up in other ways. A clash of cultures was taking place in Rome during this period, similar to the clash that would begin to take place in post-sixties America. Among other now-lost poets of antiquity, flourishing in the two decades before Julius Caesar was killed, was a rare survivor, an ancient Roman poet who can usefully be compared to Dylan, the avant-garde lyric poet Catullus. He died young (c. 54 BC) after creating a body of work that electrified Roman readers, reflected the turmoil and the modernity of Roman times, and changed the course of literary history.

Catullus has long been one of my favorite poets. For me, no other poet, except maybe Dylan, has been able to convey a sense of the pain caused by the loss of love as intensely as Catullus. Dylan wouldn't begin to make creative use of the poetry of ancient Greece and Rome until the albums he released in the twenty-first century, even though he had long been living in the Rome of his memory and imagination.

In his 2007 movie, *I'm Not There,* director Todd Haynes used Dylan's 1966 song "I Want You" for a scene in which Heath Ledger and Charlotte Gainsbourg, playing the roles of Robbie and Claire, immediately recognizable versions of Bob and Sara Lownds, first fall in love. The song encapsulates first love, joyous, and just right for that moment, with its highly poetic verses and its simple, direct refrain: "I want you, I want you / I want you so bad / Honey, I want you." Catullus too captured in his poetry the first flush of love, for instance in one of his "kiss" poems: "Suns can set and then come back again, / When our short day sets once and for all, / our night must be forever to be slept. / Give me a thousand kisses, then a hundred, / then another thousand and second hundred, / then still another thousand, then a hundred."

But the lyrics of Catullus and of Dylan mostly share a focus on love that is lost, that doesn't work out—that's where the poetry is. So, for instance, Catullus Poem 11, one of his last poems to Lesbia, the name he gave to the Muse (recalling Sappho, who lived on the Greek island of Lesbos), who inspired his love song. He begins with an address to two acquaintances, whose task it will be to take a message to Lesbia: "You who are ready to try out / whatever the will of the gods will bring / Take a brief message to my old girlfriend / words that she won't like. / Let her live and be well with her three hundred lovers, / Not really truly loving them / but screwing them all again and again." The poem ends by shifting the brutal tone and bringing out the hurt and the love that is still there: "Let her not look back for my love as before / which through her fault has fallen like a flower on the edge of a meadow / nicked by the blade of a passing plough."

By 1975, whatever the realities of his relationship with his wife, Sara, Dylan was, like Catullus as time went by, approaching the end of a relationship in trouble, and he constructed a lyric voice that made art from that situation. The song we already saw, "If You See Her, Say Hello," is similarly about a relationship that is over:

If you see her, say hello, she might be in Tangier
She left here last early spring, is livin' there, I hear
Say for me that I'm all right though things get kind of slow
She might think that I've forgotten her, don't tell her it isn't so

The song, separated from the autobiographical, is like Catullus's poem, and is there for anyone who has shared that loss and hurt. Like Catullus, Dylan too imagines the rival who has supplanted him: "If you're making love to her . . ." Back in Ann Arbor, I was reading the Latin poetry of one, and listening to the songs of the other. And that is how Catullus and Dylan, both lyric poets, sharing common human situations across twenty centuries, have become inextricably linked in my mind, and why they belong together.

Catullus would have been much more familiar in America in the early 1960s, as is clear from an early scene from *Cleopatra*. It was the highest-grossing film of 1963, won four Academy Awards, and still lost money, so costly was its production. It is highly likely that Dylan, like millions in America and around the world, saw it that year, as I did back in New Zealand. Elizabeth Taylor's Cleopatra, kittenish and scantily clad on her couch in Alexandria, receives a visit from Rex Harrison's Julius Caesar. Richard Burton's Mark Antony is waiting in the wings, and will take over after the assassination of Caesar on those Ides of March. Her spies have reported on Caesar's movements:

CLEOPATRA: This morning early, you paid a formal visit to the tomb of Alexander. You remained alone beside his sarcophagus for some time. . . . And then you cried. Why did you cry, Caesar?

CAESAR, CHANGING THE SUBJECT: That man recites beautifully. Is he blind?

AN ATTENDANT: Don't you hurt him.

CAESAR: I won't. Not anyone who speaks Catullus so well.

CLEOPATRA: Catullus doesn't approve of you. Why haven't you had him killed?

CAESAR: Because I approve of *him*.

CAESAR, TO THE YOUNG SINGER, HIS WORDS MEANT FOR CLEOPATRA:
Young man, do you know this of Catullus?

Give me a thousand and a thousand kisses

When we have many thousands more,
we will scramble them to get the score,
So envy will not know how high the count
And cast its evil eye.

Several scenes later, once Cupid's work is done and Caesar and Cleopatra are lovers, she lies back on her bed and volunteers, "I've been reading your commentaries, about your campaigns in Gaul." He, skeptical: "And does my writing compare with Catullus?" She, suggestively: "Well, it's [slight pause] different?" "Duller?" he asks. "Well, perhaps a little too much description."

Unlike today's audiences, those watching the film in 1963, including Dylan, would have gotten these references. Ancient Rome and its spoken language, Latin, the biggest language club at Hibbing High and elsewhere, used to be much more relevant. As late as January 28, 1974, the cover of *Newsweek* could show Richard Nixon, H. R. Haldeman, and Rosemary Woods encoiled by the Watergate tapes in an image that was a clear allusion to the twin snakes in Virgil's *Aeneid* that devour the Trojan priest Laocoön, who is trying to urge his people not to bring the Greeks' fateful horse into the city. Readers of *Newsweek,* Dylan included, would have gotten it, either from their knowledge of Virgil or of the ancient statue of the scene, now in the Vatican. Until 1928, enrollments in Latin language courses in the United States outstripped all other languages combined. Spanish took over as the years went by, but in 1962 there were still

702,000 students studying the ancient language. Sputnik, the Cold War, and the perceived need for more science and practicality in U.S. school curricula put an end to all that. The decline began when the National Defense Education Act of 1958 omitted Latin from the curriculum—a year after Bob Zimmerman had been in Latin class at Hibbing High. It took some time to see the full effects of that measure, but by 1976 the number of Latin students had dropped sharply to 150,000, helped by the difficult nature of the language, along with its association with the church, discipline, and authority. Latin hardly fit the ethos of the counterculture.

The paradox here is that Catullus's poetry is in fact completely modern in the themes and sentiments it expresses. Those who understand his work read it for the beauty and the music of his verse, for the intensity of the personal voice, and for solace when they have loved and lost. Catullus was among the most-read poets of a number of the Beat poets. Alfred, Lord Tennyson, laureate poet of Victorian England, visiting the ruins of Catullus's house on Lake Garda in northern Italy, thought of Catullus's poem to his dead brother: "Came that 'Ave atque vale' [hail and farewell] of the poet's hopeless woe / Tenderest of Roman poets nineteen hundred years ago." The historian and politician Thomas Babington Macaulay (1800–1859) could not read Catullus's Poem 8 without weeping. It has been a favorite since Thomas Campion, the poet, musician, and doctor, translated it and put it to music in the early seventeenth century.

Unlike many in our age, Campion obviously saw no distinction between poem and song. The poem is a self-address, urging strength and resolve, after the loss of Lesbia's love:

> Poor Catullus, you should stop being a fool!
> Should realize what you see is lost is gone for good.
> Bright were the suns that once shone once for you
> When you would go wherever she would lead you.
> That girl loved as no other will ever be.
> Many playful things happened then,
> Things you wished and she then wanted too.
> Bright indeed the suns that once shone for you.
> Now she doesn't want you. You should be the same.

The poem continues, with the poet unable to get beyond the love that is lost, as he imagines her with another: "Whom will you kiss, whose lips will you nibble." Or, as Dylan put it in refrain of the 1997 song "'Til I Fell in Love with You": "I just don't know what I'm going to do / I was all right 'til I fell in love with you." Or at the end of "Love Sick," from the same album:

> *I'm sick of love; I wish I'd never met you*
> *I'm sick of love; I'm trying to forget you*
> *Just don't know what to do*
> *I'd give anything to be with you*

This is the art of Catullus and the art of Bob Dylan, then a fifty-six-year-old songwriter, the essence of which he sums up in *Chronicles: Volume One:* "experience, observation, and imagination"—qualities he shares with the Roman poet.

Another poem of Catullus, his shortest, was translated by Abraham Cowley, English Civil War poet, in the seventeenth century:

> I hate and yet I love thee too;
> How can that be? I know not how;
> Only that so it is I know,
> And feel with torment that 'tis so.

In spirit these poems share much with the songs Dylan was writing in the second half of 1962, when he was wasting away in the Village, pining for the absent Suze Rotolo, and producing some of his best work *because of* that absence. Perhaps he even knew the Catullus poem above—Miss Walker may have shown it to the Latin class, given its simplicity and brevity—as we seem to hear its echoes in a letter he wrote to Suze in 1962:

> It's just that I'm hating time—I'm trying to . . . bend it and twist it with gritting teeth and burning eyes—I hate I love you.

The songs of this period come across as heartfelt, and reflect a reality, but like the poems of Catullus, they come into being and endure through the artistry with which they capture the human condition. The connection between the lyric genius of these two poets may be coincidental, but Dylan's interest in the city in which Catullus lived, loved, lost, and died young is a very real thing.

DYLAN VISITS ROME

Bob Dylan would pay the first of many visits to Rome, also his first time in Europe, in January 1963, a side trip after performing in a BBC film in London the month before, during what was also his first trip to England. The summer before these trips, in June 1962, Suze Rotolo, Dylan's girlfriend and Muse of those years, had been taken off to Italy by her mother. Mary Rotolo disapproved of her young daughter's relationship with Dylan, and Suze herself was troubled by the stress that Dylan's exploding fame was beginning to cause. Originally scheduled to return by Labor Day, she stayed on past the summer, studying art for the rest of the year in the Umbrian city of Perugia. But Dylan's trip to Rome had nothing to do with retrieving Suze, who by then had returned to New York. So why did he first visit Rome, and not Paris, Berlin, or Madrid? The liner notes to his second album, *The Freewheelin' Bob Dylan,* mention that he actually performed on this first trip to Rome, at the Folkstudio in the bohemian region of Trastevere ("across the Tiber"), "in its heyday a Greenwich Village–style club with three or four performers every night and a generous open-stage policy." It seems likely that Rome and its fascination had existed in Dylan's imagination, dating back just a few years before the trip to his study of Latin and the Latin Club, all those movies, and his stage debut as a Roman soldier, with the highlights of the eternal city, not least of all its Colosseum (or "Coliseum") and gladiators, appealing to his young mind.

Dylan's separation from Suze Rotolo gave us some of his greatest songs, written while they were apart: "Don't Think Twice, It's All Right," "Tomorrow Is a Long Time," "One Too Many Mornings," "Girl of the North Country," and of course, "Boots of Spanish Leather," its first six verses a dialogue between the singer and his lover. Dylan and Rotolo had corresponded during her absence, and the seventh verse of the song captures the pain of the man who has been left behind:

I got a letter on a lonesome day
It was from her ship a-sailin'
Saying I don't know when I'll be comin' back again,
It depends on how I'm a-feelin'

Dylan and Suze would later get back together, but written during those first days that Dylan spent in Rome, it preserves the evidence of a painful memory of separation, "across that lonesome ocean." Dylan would sing the song on Studs Terkel's show in May 1963. Terkel asks for a love song. Dylan: "You wanna hear a love song?" Terkel: "Boy meets girl. Here's Bob Dylan, boy meets girl." Dylan strums a chord or two—and corrects Terkel, "Girl leaves boy."

Dylan's trip to Rome also gave us a song called "Goin' Back to Rome," which he would perform on February 8, 1963, at Gerde's Folk City, once he returned from his trip. "Goin' Back to Rome" is not copyrighted, or included among the songs on Dylan's official website, but it is preserved on the bootleg recording "The Banjo Tape," transcribed here correctly for the first time:

Hey, well, you know I'm lying
But don't look at me with scorn.
Well you know I'm lying
But don't look at me with scorn.
I'm going back to Rome
That's where I was born.

Buy me an Italian cot and carry,
Keep it for my friend.
Buy me an Italian cot and carry

Keep it for my friend.
Go talk to Italy
All around its bend.

You can keep Madison Square Garden
Give me the Coliseum.
You can keep Madison Square Garden
Give me the Coliseum.
So I don't wanna see the gladiators
Man I can always see 'em.

While the lyrics here are obscure, they may not be pure nonsense. We can connect "going back to Rome" with the fact that the twenty-one-year-old had actually been in Rome the month before. And when he sings, "Buy me an Italian cot and carry, / Keep it for my friend," we wonder about the identity of the friend for whom he would buy the portable baby cot. Could it have been for Suze Rotolo herself, whose pregnancy later that year was terminated by an abortion? Suze was presumably in the audience at Gerde's that night, having just been reunited with a Dylan much happier than the one who had been moping around the Village in the second half of 1962 while she was off in Italy.

The song's claim of a birthplace in Rome is an early instance of Dylan's tendency to create environments for his various identities and characters. Another example is in the traditional "Man of Constant Sorrow" (1962), where the narrator is "going back to Colorado," where he was "born and partly raised." Other versions of this song have the singer "born and raised" in old Kentucky, others in San Francisco, and so on. As long as the meter of the place is the same (Cólorádo, Sán Francísco, etc.), anything works, variety and change of place and time being a natural feature of folk songs. But for Dylan, Rome is different. It is hard not to connect his staking a claim for his birthplace in Rome with other utterances that try to create a new point of origin for himself, one that makes more sense in the creative mind of this genius, like this moment in his 2004 *60 Minutes* interview with Ed Bradley. Questioned about changing his name from Robert Zimmerman to Bob Dylan, he replied: "You're born, you know, the wrong names, the wrong parents."

Even before the trip to Rome and the penning of the song "Goin' Back to Rome," which followed the visit, was an even earlier song, "Long Ago, Far Away," sung before he would have known of the Folkstudio in Rome, in Minneapolis friend Tony Glover's apartment on August 11, 1962. Recorded in November 1962, the song shows he was thinking of ancient Roman times. It shows a debt to gospel, as it considers human cruelty throughout history, from the point of view of those who suffer, not least the crucified Jesus, with whom the song begins and ends:

To preach of peace and brotherhood
Oh, what might be the cost?
A man he did it long ago
And they hung him on a cross.

Then the ironic, even sarcastic, refrain, implying that nothing has changed:

Long ago, far away,
Things like that don't happen
No more, nowadays

The thrust of the song is along the lines of Woody Guthrie's song "Jesus Christ":

This song was made in New York City
Of rich man and preachers and slaves
If Jesus was to preach like He preached in Galilee
They would put Jesus Christ in His grave.

Of the examples in the five verses in the body of Dylan's song, only two specify historical moments, the chains of slaves "during Lincoln's time," and in the striking image of the second to last verse, absent from the official Dylan website:

Gladiators killed themselves
It was during the Roman times
People cheered with bloodshot grins
As eyes and minds went blind
Long ago, far away
Things like that don't happen
No more, nowadays

"Goin' Back to Rome" has "always see 'em" rhyming with "Coliseum," and we will see this rhyme used again, a few years later, in a famous and much better song from 1971, "When I Paint My Masterpiece." The first two verses of that song are all about Rome. It is thought to preserve the memory of another trip Dylan took to Rome, following the 1965 tour of England that was the subject of D. A. Pennebaker's film *Don't Look Back*. According to this plausible theory, Dylan went back in the company of his new Muse, Sara Lownds, whom he would marry by the end of the year. Sara left her husband, Hans Lownds, to take up with Dylan, and it is hard not to connect this reality with one of Dylan's most iconic lines, from "Idiot Wind," written in 1974: "They say I shot a man named Gray and took his wife to Italy."

Sara is absent in name from "When I Paint My Masterpiece," thus allowing the singer in the scene Dylan paints to have an assignation with "a pretty little girl from Greece," or in the official lyrics, "Botticelli's niece," who will be able to help out with painting the masterpiece in the title and in the last line of each verse:

Oh, the streets of Rome are filled with rubble
Ancient footprints are everywhere
You can almost think that you're seein' double
On a cold, dark night on the Spanish Stairs
Got to hurry on back to my hotel room
Where I've got me a date with Botticelli's niece
She promised that she'd be right there with me
When I paint my masterpiece

It is worth noting that "When I Paint My Masterpiece" was the regular opener for the first of the two Rolling Thunder Revue tours in the fall of 1975, which also featured the new song "Sara." "Sara" is pretty much the opposite of "Idiot Wind" in its lyrical and sweet memories of their decade-long relationship, children and all. The only song with a title and lyrics that are unambiguous on the identity of the lover, its last two choruses end with a hint of the breakup toward which the two were headed: "You must forgive me my unworthiness. . . . Don't ever leave me, don't ever go." By this time, Dylan was involved with other women, and when the 1976 part of the Rolling Thunder Revue resumed in Lakeland, Florida, on April 18, 1976, "Sara" was gone from the setlist, its position taken over by "Idiot Wind"—the first public performance of the song that had come out fifteen months earlier. In a televised performance on May 23, 1976, the eve of Dylan's thirty-fifth birthday, he delivers a glorious, impassioned version of it with his wife, children, and mother in the audience. It was likely not much appreciated by Sara, sitting there, the object of much of the song's anger and venom. Their divorce was finalized on June 29, 1977.

Clearly Dylan feels a connection to the antiquity of Rome, as he does with no other place. When he first traveled there in 1963, he was immediately inspired. That's why in January 1963 he would write and sing the words "Goin' back to Rome, / That's where I was born." That trip to Rome, subsequent trips, and the adoption of the city as the place where he was born seemed years later to incite a kind of artistic rebirth for Dylan, or at least it coincided with that rebirth.

PRESS CONFERENCE IN ROME, 2001

Bob Dylan would be back in Rome on July 23, 2001, for an interview with a group of European journalists who had been listening that morning to an advance copy of his new album, *"Love and Theft,"* due out the following September. This was in between concerts in Pescara, on the Adriatic coast of Italy, and Anzio, on the coast south of Rome. On this tour he performed in Scandinavia, Germany, England, Ireland, but it was Rome that he chose for the press conference that would plant a few clues about the new directions in his songwriting. One of the reporters asked an early question:

Are you enjoying to be in Rome?

Oh yeah.

You're often here in Rome.

Pretty regular huh?

You write songs about [Rome].

Quite a few.

It is interesting that Dylan doesn't limit himself to just the obvious song or to any one song, but "quite a few." The journalist misses his observation, and is just thinking of the song in question:

"Paint My Masterpiece."

Exactly.

You speak exactly of this here. . . .

Exactly. This is it . . . Spanish Steps.

The Hotel de la Ville, where the interview took place, was a little way along the Via Sistina from the top of the Spanish Steps—another prop for the interview. As the press conference continued, Dylan proceeded to lay down a trail for journalists to follow:

My songs are all singable. They're current. Something doesn't have to just drop out of the air yesterday to be current. You know, this is the Iron Age, we're living

> in the Iron Age. But, what was the last age, the Age of Bronze or something? We can still feel that age. I mean if you walk around in this city, you know, people today can't build what you see out there. Well at least, you know when you walk around a town like this, you know that people were here before you and they were probably on a much higher, grander level than any of us are. I mean it would just have to be. We couldn't conceive of building these kind of things. America doesn't really have stuff like this.

This looks close to being scripted, preplanned, and he gets back to it later in the interview after the journalists fail to pick up on where he was going:

> We've talked about these ages before. You've got the Golden Age, which I guess would be the age of Homer, then we've got the Silver Age, then you've got the Bronze Age. I think you have the Heroic Age someplace in there. Then we're living in what some people call the Iron Age, but it could really be the Stone Age. We could really be living in the Stone Ages.

Dylan's language was tantalizing and now caught the attention of those present. After the first of these comments, where he said that something can be "current," but also as old as the Age of Bronze, and "we can still feel that age," one of the journalists sensed an opening:

> Do you read history books?
>
> *Huh?*
>
> Do you read books about history? Are you interested about that?
>
> *{pause} Not any more than just would be natural to do.*

Earlier in the interview, he had been asked about "new" poets on *"Love and Theft."* His response deflected the truth, typical of Dylan, for whom there were lots of new poets beginning to enter his arsenal:

> Are you still eagerly looking for poets that you may not have heard of or read yet? Or do you go back to the ones that have interested you like maybe Rimbaud? [*pause, followed by a sigh of sorts*]
>
> *You know I don't really study poetry.*

He may not study poetry, but the ancient footsteps that are everywhere on "When I Paint My Masterpiece" are also on "Lonesome Day Blues," one of the songs from the new album that the reporters had just been listening to, and the one that echoes the lines he had adapted from the Roman poet Virgil:

I'm gonna spare the defeated—I'm gonna speak to the crowd
I'm gonna spare the defeated, boys, I'm gonna speak to the crowd
I'm goin' to teach peace to the conquered
I'm gonna tame the proud.

Dylan, however, was not going to spell things out more than he had already done. That's not his style. The journalists would have to make that connection for themselves. The interview ended with applause from the twelve satisfied and grateful reporters. "Now I'm gonna go see the Colosseum," he told them. In reality, this was a highly unlikely proposition, though a drive-by could have happened. In 1965 he and Sara could just have pulled off a visit, but in 2001 Dylan would have been mobbed in such a public and open space. As is often the case with Dylan, he was visiting the song in that moment and in his mind's eye.

The second verse of "When I Paint My Masterpiece" has the singer in the Colosseum, reusing the rhyme he had come up with for "Goin' Back to Rome":

Oh, the hours I've spent inside the Coliseum
Dodging lions and wastin' time
Oh, those mighty kings of the jungle I could hardly stand to see 'em

Yes, it sure has been a long, hard climb
Train wheels runnin' through the back of my memory
When I ran on the hilltop following a pack of wild geese
Someday, everything is gonna be smooth like a rhapsody
When I paint my masterpiece

With the fourth line—"It sure has been a long, hard climb"—he looks across the ten short years in which so much had happened since he had set out from his native Minnesota and arrived in Greenwich Village. But what about the next lines? The train wheels running through the back of his memory might seem to take us into another Dylan song, "Bob Dylan's Dream," from 1963, when he falls asleep while "riding on a train goin' west" and is taken back to the days of his youth, and to the first few friends he had back then—by way of a nineteenth-century folk song.

But why does his memory train have him "running on the hilltop following a pack of wild geese"? Clinton Heylin tried to make sense of it: "he has returned in his time machine to Hibbing, remembering a time when he 'ran on a hilltop following a pack of wild geese.'" But it is hard to find space for Hibbing in this song, whose next verse, "I left Rome and landed in Brussels," would sandwich his hometown on a short plane ride between the capitals of Italy and Belgium. And following geese in Hibbing doesn't make too much sense, unless the geese were in some classroom at Hibbing High, either during a Latin class or in discussion at the Latin Club. The wild-goose chase to which his memory goes back from the streets of Rome is more likely to refer to one of the favorite stories about ancient Rome, bound to have been on the quiz shows of the Latin Club, in which the sacred geese of the goddess Juno on the Capitoline Hill warned the Romans that invading tribesmen from Gaul were attacking the religious center of Rome. Virgil has the scene, along with other high points of Roman history, Romulus and Remus, the Sabine women, and Tarquin the Proud, on the shield that the hero Aeneas, Rome's founder, carries into battle in the *Aeneid:*

> And here the silver goose was fluttering
> Through gilded porticos cackling that the Gauls were
> at the gate

"CHANGING OF THE GUARDS" AND THE SOULS OF THE PAST

It would be some years before the streets of Rome, or at least some things Roman, came back into his lyrics, but in one song from the immediate pre-Christian—in some ways not even pre-Christian—phase he can be seen reaching back through the years and the centuries, giving us fragments of worlds, hard to unravel or pin down, but highly evocative. "Changing of the Guards" was put out as the first track of the 1978 album *Street Legal*. The opening words of the song, and the album, have generally been seen as taking stock, looking back across the years to the beginning of his career in 1962: "Sixteen years / Sixteen banners united over the field." Asked about these numbers in an interview with Jonathan Cott in November 1978, Dylan—of course—denied the relevance of the math, as he denies any single meaning for his songs. The images, situations, and characters that this song rolls out put it almost beyond overall interpretive reach—"It means something different every time I sing it." The song proceeds through an array of figures, across fields with the good shepherd grieving, desperate men and desperate women, perhaps the music industry with which he had been dealing in those years: "Merchants and thieves, hungry for power, my last deal gone down," and later, "Gentlemen he said / I don't need your organization, I've shined your shoes / I've moved your mountains and marked your cards." Just as it can mean something different every time he sings it, so it can mean something different every time I hear him sing it, depending on what images, all generally mysterious, are flashing by.

To me the song has always belonged in the world of Rome, as well as in the Christian world to which the album so clearly points. The "good shepherd," "angels' voices whisper to the souls of previous times," "Eden is burning"—the Christian aspects are obvious, and so are the ancient Roman ones: Dylan's conversation with Cott at one point turned to the antiquity of the song: *Those lines seem to go back a thousand years into the past.* Dylan agrees, and he singles out "Changing of the Guards," but takes it back further—a thousand years only gets you halfway where you need to go, so he corrects himself:

> They do. "Changing of the Guards" is a thousand years old . . . [it] might be a song that has been there for thousands of years, sailing around in the mist, and one day I just tuned in to it.

Thousands of years get you back to the Roman gods, present in the fourth verse:

> *They shaved her head*
> *She was torn between Jupiter and Apollo*
> *A messenger arrived with a black nightingale*
> *I seen her on the stairs and I couldn't help but follow*
> *Follow her down past the fountain where they lifted*
> *her veil*

What, we might wonder, is this song doing with Jupiter, king of the Olympian gods, and Apollo, god of prophecy and music, again returning to the "questions on Roman gods and goddesses" for the Latin Club radio show? One god is the figure of ultimate authority, the other the divine musician, expert on the lyre—*cithara* (again = "guitar"). But it goes deeper than that.

On the page facing the text of the song in the official lyrics books *Bob Dylan: Lyrics: 1961–2001* (2004) and *Bob Dylan: The Lyrics: 1961–2012* (2016) there is the same illustration, a grainy black-and-white version of part of the cover of *Street Legal,* Dylan standing at the foot of some stairs, one hand on hip, looking up the street. Superimposed on the image is a telltale sign, a typewritten draft of eight lines of the song, a draft of the fourth verse and a very preliminary, barely recognizable draft of the eighth with pencil or pen corrections (here in italics). For neither verse is there any trace of the short, opening line that so marks the song musically: "They shaved her head. . . . I stumbled to my feet. . . ." Here is how our draft verse, five full lines long, started out:

> *I stared into the eyes—**Ages roll**—upon **Jupiter** and **Apollo***
> ***Midwives** stroll between **jupiter** and **apollo***
> *Struggling **babes** past (Between the sheets of . . . **Destiny's***
> *faces*
> ***miraculous** one-eyed glory*

Almost every word from the draft just quoted could find a home and a source in Virgil's messianic poem *Eclogue* 4, a poem from around 40 BC about the ages Dylan would later talk about in the Rome conference: The Sibyl, prophet of (a) Apollo, predicted the birth of a (b) miraculous (c) baby, to be the offshoot of (d) Jupiter, helped into the world by Roman (e) midwife goddess Lucina. The Fates with the approval of (f) Destiny give the order that the (g) "Ages run." At the birth of the child, Virgil predicts the "Iron Age will end and the Golden Race will spring up," a reversal of the process Dylan would later talk of in the Rome conference. Only Jupiter and Apollo survived from Dylan's draft into the final version of the song, whose own messianic vision looks to the Christian images the singer would embrace the next year with the 1979 album *Slow Train Coming:*

> *But Eden is burning, either get ready for elimination*
> *Or else your hearts must have the courage for the changing of the guards.*

But the traces of Virgil, and the beginnings of a textual engagement with the poetry of Greece and Rome, are to be found here in the "song that has been there for thousands of years," as Dylan put it.

TEMPEST: GOIN' BACK TO ROME—AGAIN

Bob Dylan chose September 11, 2012, as the U.S. release date for what is now his latest original album, and one of his best: the critically acclaimed *Tempest*. The worlds created in the songs on this album come out of the song tradition, reading, and still fertile and exuberant imagination that Dylan has been drawing from for years. Those worlds, created out of his experience, observation, and particularly his imagination, resist any sort of easy definition. They are strange and beautiful, ominous and dangerous, and utterly compelling. The characters in "Soon After Midnight" come thick and fast: "a gal named Honey" took his money; "Charlotte's a harlot, dresses in scarlet / Mary dresses in green," while he's "got a date with the fairy queen." The identities are impenetrable, but if you enter into the world of this song, beautiful in performance, where it has been a favorite, that is a sufficient world, its final line leaving everything hanging there: "It's soon after midnight and I don't want anyone but you." Or take "Tin Angel," the album's eighth track. For no apparent reason, it shares its title with that of a Joni Mitchell song, and it also starts out with the exact first line of Woody Guthrie's "Gypsy Davy"—"It was late last night when the boss came home." It then turns into a twenty-eight-verse ballad of narrative and dramatic dialogue that ends up in a cross between the death scenes of *Othello* and *Hamlet,* with the boss, his wife, and his rival, Henry Lee, "chief of the clan" but otherwise unidentified, all dead, "together in a heap."

"Tempest," the album's ninth track, is even longer, a forty-five-verse ballad about the sinking of the *Titanic.* The tune and the first two verses are from the song "Titanic" by the Carter Family; the rest is from the highly cinematic mind of Dylan. The song includes lines from another Roman poet: in Dylan, "Davey the brothel keeper / Came out, dismissed his girls"; in Juvenal, "the pimp was already dismissing his girls" (Juvenal

6.127). Juvenal turns up on other tracks of *Tempest,* confirmation that Dylan is drawing from the Roman satirist. "Tempest" is crowded with scenes of chaos, civil war even breaking out at one point: "Brother rose up against brother / In every circumstance / They fought and slaughtered each other / In a deadly dance." In "Bob Dylan Unleashed," Mikal Gilmore's September 27, 2012, interview with Dylan about *Tempest,* Dylan describes it as a record "where anything goes and you just gotta believe it will make sense."

In the same interview, talking of the changes he and his music had gone through over the years, Dylan offered an interesting detail: "I went to a library in Rome and I found a book about transfiguration." We'll return to that library, and to transfiguration. For now Rome "transfigured" will play an important role in two songs on the album, whose components have echoes of the Roman identities of some of the earlier songs.

"SCARLET TOWN"

"In Scarlet Town, where I was born," the first line of the sixth song on *Tempest,* takes us back fifty years to Bob Dylan in Greenwich Village, at the Gaslight Cafe. The ten songs he sang back then were taped informally and released on *Live at the Gaslight 1962*. Anyone who doubts Dylan's ability to sing with a beautiful and melodious voice when he wants to should listen to the second to last track on that album, Dylan's version of the traditional seventeenth-century Scottish ballad "Barbara Allen."

William lies on his deathbed wasting away for the unrequited love of Barbara Allen, whom he omitted to toast down at the tavern, so giving her the mistaken impression he didn't care for her. Far from it. She is sent for and when she realizes he in fact loved her, she too dies for love of him. From his grave a red, red rose grows up, from hers a briar, the entwining of the two plants uniting them at least in death.

In Dylan's new song, "Scarlet Town," the characters familiar from the ballad don't figure until verse three, where Barbara Allen has a new name: "Scarlet Town in the month of May / Sweet William on his deathbed lay / Mistress Mary by the side of the bed / Kissing his face, heaping prayers on his head." We then enter the world of nursery rhyme as the next verse calls on Little Boy Blue to blow his horn. But otherwise ballad and nursery rhyme play no real role in the song, gone for good and absent from the last eight verses. "Uncle Tom still working for Uncle Bill" in the second verse, and the last three verses, addressed to a woman, are now in a contemporary world of regret, "A lot of things we didn't do that I wish we had."

Scarlet Town is indeed, as Dylan said of the album at large, a place "where anything goes." His 1962 "Barbara Allen" had begun "In Charlotte Town, not far from here," clearly heard in the opening of the new song—"In Scarlet Town where I was born," but it has other company, now from the unreleased song of 1963, "Goin' back to Rome / That's where I was born." Beyond this particular echo, we encounter a literary and cultural pastiche that forms the world of this song. Here "the streets [of Rome?] have names that you can't pronounce." "In Scarlet Town you fight your father's foes / Up on the hill a chilly wind blows," where the hill in this context resembles the one in "When I Paint My Masterpiece," the Capitoline Hill, one of the Seven Hills of Rome, which may lie behind another of the images in the song: "The Seven Wonders of the World are here." The father is most easily Julius Caesar, whose foes were fought and defeated by his adoptive son, the future emperor Augustus. As Dylan said in the same interview after the album came out:

> Who knows who's been transfigured and who has not? Who knows? Maybe Aristotle? Maybe he was transfigured? I can't say. Maybe Julius Caesar was transfigured.

Transfigured into this song perhaps, now dead as his son fights his father's foes?

One verse of the song, the sixth, points in its entirety to Rome and adds a new ingredient, that of Christians under Rome or in Rome:

> *On marble slabs and in fields of stone*
> *You make your humble wishes known*

I touched the garment but the hem was torn.
In Scarlet Town where I was born

"Marble slabs and in fields of stone" has an ancient world feel to it, perhaps the Roman Forum, while the next lines point toward biblical lands and to the woman in the Gospel of Luke 8.43–48 who makes her humble wishes known by touching the hem of Jesus's garment as Jesus passes by, and is immediately cured of her chronic bleeding. Dylan may even be channeling Sam Cooke's or some other version of the gospel song "Touch the Hem of His Garment." But the biblical and the Roman have always been side by side, ever since he saw *The Robe,* or stood on the stage at Hibbing High acting the role of that Roman soldier.

The next song on *Tempest* didn't seem to be concealing much, or so it appeared when the title was announced. It looked like things were headed for the eighth century BC, back to Romulus and Remus and the other kings of Rome, perhaps a lesson from the would-be Roman history teacher.

"EARLY ROMAN KINGS"

In the seventh track of *Tempest,* Rome made it into a song title, "Early Roman Kings." As it would emerge, the song at first had little to do with actual Roman kings like Romulus or Tarquin the Proud. That would be too easy:

All the early Roman Kings in their sharkskin suits
Bowties and buttons, high top boots
Driving the spikes, blazing the rails
Nailed in their coffins in top hats and tails
Fly away little bird, fly away, flap your wings
Fly by night like the early Roman Kings.

Dylan's voice is strong and sinister, matching the darkness of the lyrics and delivering a song that in performance would become his favorite from the album, played at well over three hundred concerts by 2017. The Roman Kings turned out to be a gang from the Bronx, one of many active in the 1960s and '70s, mentioned in Sol Yurick's 1965 novel, *The Warriors,* based on actual gangs that flourished at the time. But that too would be too easy to suit the songs of this album, whose worlds connect the old and the new, as Dylan put it, and which correspond to nothing objectively observable. The second verse in fact takes us back to ancient Rome:

All the early Roman kings in the early, early morn'
Coming down the mountain, distributing the corn
Speeding through the forest, racing down the track
You try to get away, they drag you back
Tomorrow is Friday, we'll see what it brings
Everybody's talking 'bout the early Roman Kings

In ancient Rome the grain supply was taken very seriously. Corn distribution was a form of dole, and there was even a magistrate responsible for the allotment of cheap or free grain to an urban populace, whose unrest in times of famine, which happened often enough, would pose a serious threat to the state. In a rather modern-sounding jab at the Roman people, the poet Juvenal, whose *Satires,* written around AD 130, have a modern ring to them, noted that the people who used to care about electing politicians and military leaders suddenly had a desire for just two things, "bread and circuses" (*Satire* 10.81). That's what Dylan's early Roman kings are up to in the second verse: "distributing the corn" and "racing down the track"—Dylan knows that the racetrack of ancient Rome is the Circus Maximus, a short walk from the Colosseum. The classical pastiche runs through the song. Juvenal comes back in the last verse, in the line "Gonna put you on trial in a Sicilian court," which Dylanologist Scott Warmuth plausibly connected to a different poem of Juvenal, *Satire* 6, where the household regime of a Roman matron is said to be as cruel as that of a Sicilian court—referring to the proverbial cruelty of the ancient tyrants of Sicily. We will return to this song later in the book, where its lyrics take the identity of the Roman kings back as far as you can go in Western literature, to Odysseus, Homer's vagabond king of the island of Ithaca. But before that, let's make one more trip back to Rome, with Dylan in the fall of 2013.

GOIN' BACK TO ROME: SETLIST SUNDOWN

The recent albums of Bob Dylan have become the essential elements of the singer's concert setlists in recent years. He reaches back selectively to the songs of the sixties and seventies, but to the annoyance of those who yearn for the old songs and the old voice, the world he constructs for performance today draws from his recent masterpieces, increasingly interwoven with some of the songs from the Great American Songbook of the mid-twentieth century, five albums' worth by 2017. The setlists have also become more fixed, with little variation, but with constant modulations in his singing and in his band's accompaniment in the performed versions of those songs.

Then something happened that is without parallel in Dylan's concerts of the last several years. In the middle of the 2013 fall concert tour, which had begun in Oslo and gotten as far as Milan, Dylan and the band headed south to Rome, for two concerts at the Atlantico, a four-thousand-seat venue in the ancient capital. For those two nights in Rome, November 6 and 7, and only those two nights, he almost completely threw away the setlist that everyone was expecting. "Bob is serving the crowd a la carte," wrote an Italian reviewer of the first Rome show. To examine the details of the menu over these two concerts is to reveal an artist who was connecting the place he first visited more than fifty years before with the art he had created across those years in over thirty-five hundred performances. The two concerts were retrospective in the extreme, with not a single song from the 2009 album *Together Through Life* or his great 2012 album, *Tempest,* which had become a concert staple.

Remarkably, for the course of these two Roman sets, Dylan performed seventeen songs that appeared in none of his other concerts that year. Even more striking is the fact that fifteen of the songs in the two Rome setlists have to date featured in no regular concert since, with one exception, to which we'll eventually get. These are some of Dylan's finest and best-loved songs, a veritable greatest-hits list: "Leopard-Skin Pill-Box Hat," "Honest with Me," "Queen Jane Approximately," "Ain't Talkin'," "Most Likely You Go Your Way (and I'll Go Mine)," "Boots of Spanish Leather," "Every Grain of Sand," "Like a Rolling Stone," "It Ain't Me, Babe," "Man in the Long Black Coat," "Positively 4th Street," "Rollin' and Tumblin'," "When the Deal Goes Down," "Under the Red Sky," and "I Don't Believe You (She Acts Like We Never Have Met)." In hindsight, four years after the Rome concerts, this list reads like a swan song offered up to the Romans who have interested Bob Dylan across the centuries.

Two of the songs he performed in Rome, "Girl of the North Country" and "Boots of Spanish Leather," bring in a further dimension. These two songs had been written or partially written in Italy in early 1963, both with rich folk music traditions behind them, and if any single woman is behind "Girl of the North Country," she probably came from where the song put her, up in the woods of northern Minnesota. But their lyrics were also inextricably linked to Dylan's relationship with Suze Rotolo, whose absence had helped generate those songs fifty years before the Rome concerts. Suze died on February 25, 2011, so perhaps Dylan's performance of "Boots of Spanish Leather" in Rome was also a tribute to her. Whether or not that is so, only one person knows, and we're unlikely to hear from him on that score. Now it's time to follow the trail back to where it all began in earnest, on the streets of New York City, in the early sixties.

"VIBRATIONS IN THE UNIVERSE": DYLAN TELLS HIS OWN STORY

YOU DON'T KNOW ABOUT ME WITHOUT THAT YOU HAVE READ A BOOK BY THE NAME OF "THE ADVENTURES OF TOM SAWYER," BUT THAT AIN'T NO MATTER. THAT BOOK WAS MADE BY MR. MARK TWAIN, AND HE TOLD THE TRUTH, MAINLY. THERE WAS THINGS WHICH HE STRETCHED, BUT MAINLY HE TOLD THE TRUTH.

—MARK TWAIN, *THE ADVENTURES OF HUCKLEBERRY FINN*

I DIDN'T KNOW WHAT AGE OF HISTORY WE WERE IN NOR WHAT THE TRUTH OF IT WAS. NOBODY BOTHERED WITH THAT. IF YOU TOLD THE TRUTH THAT WAS ALL WELL AND GOOD AND IF YOU TOLD THE UN-TRUTH, WELL THAT'S STILL WELL AND GOOD. FOLK SONGS TAUGHT ME THAT.

—BOB DYLAN, *CHRONICLES: VOLUME ONE*

Dylan's memoir, *Chronicles: Volume One,* was published on October 5, 2004, classified by his publisher as "biography." Readers soon realized they were dealing with anything but a biography in the traditional sense. Dylan's life seems almost impossible to contain or set in any category—we see only what he lets us see. The five chapters of *Chronicles* take the reader through selected periods of Dylan's life, in a style that is delightful, lively, and clear, and that filters the autobiographical through the artist's creative imagination. The book covers only a tiny fraction of the enigmatic life of Dylan. In reality, it's more like a play, in five acts, but with constant flashbacks, fast-forwarding, inventions, and falsehoods. The structure is elegant: Chapters 1, 3, and 5 read like the truth, while Chapters 2 and 4 read somewhat like fiction, and as Dylan says, "that's still well and good." Time is rarely linear, and particularly within those more imaginative second and fourth chapters a sense of the surreal predominates. Incongruities and deliberately chaotic juxtapositions abound, to hilarious and hugely pleasurable effect. This is not an autobiography in any sense of the word.

Chapter 1, "Markin' Up the Score," opens with a conversation between Dylan and Lou Levy, of the Leeds Music Publishing Company. Dylan does not give dates here or anywhere else in the book, but the scene must belong to late 1961, after John Hammond, "the great talent scout and discoverer of monumental artists," as Dylan put it, signed him on with Columbia Records, thus setting in motion the public career of the twenty-year-old. Similarly, other chapters focus on Dylan's story at other very specific moments in his career, again without any linearity, and maintaining silence on much of what Dylan fans would like most to hear about—the astonishing music from 1965 to 1966, the Rolling Thunder Revue tours of 1975–76, the Christian years of 1979–81. The fluid treatment of time in *Chronicles: Volume One* is reminiscent of his practice in many of his songs, a deliberate frustrating of attempts to impose temporal order on his world.

Instead, Chapter 3, "New Morning," sharing its title with Dylan's 1970 album, starts with the twenty-seven-year-old Dylan returning from Hibbing to New York and reuniting with his wife and children—he never names them, and in the real world "my wife" in Chapter 3, Sara, is a different human being from "my wife" in Chapter 4, Carolyn Dennis. The year must be 1968, since he had gone back to Minnesota that year for the funeral of his father, Abe Zimmerman, who died on June 5. A prominent theme of the chapter is a songwriting collaboration with poet and playwright Archibald MacLeish, also librarian of Congress under FDR. Dylan and MacLeish didn't quite see eye to eye on the songs Dylan had delivered, and Dylan ultimately realized the project wasn't going to work, but their encounter gives him the opportunity for some fine writing, descriptive vignettes of "Archie's" place in rural Conway, Massachusetts, with lively conversation between the two. At one point, MacLeish asks Dylan if he has read Sappho and Socrates (meaning Plato, presumably, since Socrates left no writing). "I said, nope I hadn't and then he asked me about Dante and Donne. I said not much." The talk then turns to a topic familiar from the

Rome press conference of 2001 and the draft of "Changing of the Guard" (111):

> MacLeish tells me that he considers me a serious poet and that my work would be a touchstone for generations after me, that I was a postwar Iron Age poet but that I had seemingly inherited something metaphysical from a bygone era.

Their conversation, or at least Dylan's reconstruction of it, was heading back in time, and soon enough he had returned to Homer, the father of Western literature, the poet whom in the Rome press conference of 2001 Dylan had labeled as belonging to the Golden Age (112): "MacLeish tells me that Homer, who wrote the *Iliad,* was a blind balladeer, and that his name means 'hostage'"—both details true enough. Whether or not Dylan was recalling a conversation from more than thirty years earlier, the presence of Homer has more to do with his literary and artistic activities of 2004.

Chapter 4, "Oh Mercy," jumps to 1989, in another finely crafted section of writing, much of it clearly made up. In the May 23, 2011, issue of *Rolling Stone,* writer Andy Greene quotes critic Clinton Heylin's reaction to this chapter:

> "As far as I can tell almost everything in the *Oh Mercy* section of *Chronicles* is a work of fiction," Dylan

> biographer Clinton Heylin recently said. "I enjoy *Chronicles* as a work of literature, but it has as much basis in reality as {Dylan's 2003 film} *Masked and Anonymous*, and why shouldn't it? He's not the first guy to write a biography that's a pack of lies."

The book is best seen as reality filtered through the three ingredients of creativity, in the memoir as in his songwriting: "creativity has much to do with experience, observation, and imagination." Take, for example, his descriptions of singer Fred Neil, who ran the daytime performances at the Café Wha? in Greenwich Village, where Dylan was a regular after he arrived in New York City in January 1961. In a 1984 interview, Dylan described Neil: "Fred was from Florida, I think . . . and he had a strong, powerful voice, almost a bass voice. And a powerful sense of rhythm." In *Chronicles,* Neil has become something else, something larger than life:

> He played a big dreadnought guitar, lot of percussion in his playing, piercing driving rhythm—a one-man band, a kick in the head singing voice. He did fierce versions of hybrid chain gang songs and whomped the audience into a frenzy. I'd heard stuff about him, that he was an errant sailor, harbored a skiff in Florida, was an underground cop, had hooker friends and a shadowy past. He'd come up to Nashville, drop off

> songs that he wrote and then head for New York
> where he'd lay low, wait for something to blow over
> and fill up his pockets with wampum.

Dylan writes of how Neil let him treat himself to "all the French fries and hamburgers I could eat at the Café Wha?" and he gives a similarly penetrating sketch of Norbert the cook, who used to leave a "greasy hamburger" for Dylan and falsetto singer Tiny Tim, who also performed at the club:

> Norbert was a trip. He wore a tomato-stained apron, had a fleshy, hard-bitten face, bulging cheeks, scars on his face like the marks of claws—thought of himself as a lady's man, saving his money so he could go to Verona in Italy and visit the tomb of Romeo and Juliet. The kitchen was like a cave bored into the side of a cliff.

Dylan's imagination and his creative, vigorous writing again and again transform these figures. Most of them were dead by the time *Chronicles* was published, so Dylan has a free hand, turning the Village into a carnivalesque museum, capturing that moment when the fifties turned into the sixties and the Beats gave way to folksingers. In the first two chapters of the book, Dylan digs these characters up and gives them new, tragic, heroic, surreal roles. The effect is not unlike that in the great song "Desolation Row": "Einstein disguised as Robin Hood" off "sniffing drainpipes and reciting the alphabet"—and all the other masks and disguises that song creates. Dylan's memory of detail is clearly formidable, but so are his imaginative and creative powers and his way with words.

THE LOST LAND

Which brings us to what *may* be his greatest creation in the book, the biggest whopper of them all. Dylan's flair for language and literature brings us to two interesting characters, Ray Gooch and Chloe Kiel. It's hard to know what to make of this remarkable couple, who appear in the second chapter, a section of the book that conjures up myth and fiction, a world outside its historical time, but one that in Dylan's "autobiography" is set in that historically frigid 1961 winter in New York City. The opening line of this chapter, "I sat up in bed and looked around," faintly recalls the opening of "Tangled Up in Blue": "Early one morning the sun was shinin' / I was layin' in bed." Only in the book, "it was midafternoon, and both Ray and Chloe were gone" (26). Night owl Dylan is a newcomer to New York, and he is crashing with Ray and Chloe, a colorful couple who some readers and reviewers believe are fictional, although rumor has it that Dylan, when questioned about them, has answered that they are real—for what that is worth, and not that it matters.

Why did Dylan call Chapter 2 "The Lost Land"? In part I suspect because for everyone the land of our youth is indeed lost, gone by the time we reach a certain age, along with the people who inhabited that land. So a land lost in time makes sense for a chapter recollecting those days, more than forty years before the book came out. But the phrase conjures up other possibilities, the mythic lost lands and utopian places set off from the world, the city of Atlantis beneath the sea, the Tibetan valley of Shangri-la, El Dorado, city of gold. Then there is *Land of the Lost,* the TV series that Dylan's children could have watched when it ran from 1974 to 1976 as their parents' marriage was falling apart. The show's dinosaurs, lizard-men, and friendly primates might have provided a welcome distraction. Whatever the meanings of the lost land, it is a memorable land, to say the least.

Dylan points us to this tradition, underscoring the mythical, when he talks about the writers whose stories attracted him even before he started plumbing the rich depths of folk song (39):

> In the past I'd never been that keen on books and writers but I liked stories. Stories by Edgar Rice Burroughs, who wrote about mythical Africa—Luke Short, the mythical Western tales—Jules Verne—H. G. Wells. Those were my favorites but that was before I discovered the folksingers.

These authors are the early-twentieth-century creators of various lost and nonexistent lands, the lost world of Tarzan, mythical places of *Journey to the Center of the Earth* and *Twenty*

Thousand Leagues Under the Sea, The Island of Doctor Moreau. The title of Chapter 2 of *Chronicles* perhaps puts us on notice about the genre we are about to read. Fact or fiction, or something in between?

Ray and Chloe's apartment, in which Dylan wakes up in the opening words of "The Lost Land," is said to be in Tribeca, in a building "near Vestry Street below Canal," close to the Hudson River. This will allow Dylan to walk over to the window and look out "into the white, gray streets and over towards the river." Dylan also describes the apartment as being on the same block as the Bull's Head, "a cellar tavern where John Wilkes Booth, the American Brutus [again those Ides of March via the chief assassin of Julius Caesar], used to drink." Dylan claims to have seen Booth's ghost in the mirror at the tavern. In reality, the Bull's Head, now closed, was well north of Canal, almost three miles away from Vestry Street, at 295 East Avenue. It would seem that Dylan needed the tavern to be close to the apartment for the purposes of the story and its drama. This too could be a lot of bull, but it makes for good reading.

Aside from giving Dylan a place to stay, Ray and Chloe seem to have had no real part in the life that Dylan was leading or people with whom he was associated in the Village, but they come alive through Dylan's descriptive language and dramatization. Dylan describes Ray as being "like a character out of some of the songs I'd been singing," ten years older than Dylan—making him all of thirty. Dylan writes (26):

> he was like an old wolf, gaunt and battle-scarred—came from a long line of ancestry made up of bishops, generals, even a colonial governor. He was a nonconformist, a nonintegrator and a Southern nationalist. He and Chloe lived in the place like they were hiding out. . . .

His "nonintegrator" status is picked up on later in the chapter when Dylan talks about the preaching of the man who had "been 'expelled with gratitude' from Wake Forest Divinity School" (27, 77):

> Ray was a Southerner and made no bones about it but he would have been antislavery as much as he would have been antiunion. "Slavery should have been outlawed from the start," he said. "It was diabolical. Slave power makes it impossible for free workers to make a decent living—it had to be destroyed." Ray was pragmatic. Sometimes it was as if he had no heart or soul.

And later:

> He wasn't somebody that would leave any footprints in the sand of time, but there was something special about him. He had blood in his eyes, the face of a man

> who could do no wrong—total lack of viciousness or wickedness or even sinfulness in his face. He seemed like a man who could conquer and command any time he wished to. Ray was mysterious as hell.

According to Dylan, they both had jobs, Ray in a tool-and-die factory in Brooklyn, Chloe as a "hatcheck girl at the Egyptian Gardens, a belly-dancing dinner place on 8th Avenue" (26). She too sounds like someone who could have been in a Dylan song: "Chloe had red-gold hair, hazel eyes, an illegible smile, face like a doll, and an even better figure," a little reminiscent of Ruby in the 1986 song written with Sam Shepard, "Brownsville Girl": "Ruby was in the backyard hanging clothes, she had her red hair tied back. . . . Brownsville girl with your Brownsville curls / Teeth like pearls shining like the moon above." This is in sync with the narrating voice of this chapter of *Chronicles,* a voice that like its author seems to belong back in the sixties, or even further back in time, maybe the forties, right out of a Raymond Chandler novel.

Ray and Chloe are often away from the apartment, which allows Dylan to take us on tours of the various parts of the place, with "five or six" thematically populated rooms. The detail is exquisite, suggestive either of a photographic memory, unfading across all the years, or something else, namely fiction—or a combination of fact and fiction. As with his songwriting, anything goes.

THE GUN ROOM

Dylan describes one room in the apartment as being full of guns:

> There were different parts of guns—of pistols, large frame, small frame, Taurus Tracker pistol, a pocket pistol, trigger guards, everything like in a compost heap—altered guns . . . guns with shortened barrels, different brands of guns—Ruger, Browning, a single-action Navy pistol, everything poised to work, shined out. You'd walk into this room and feel like you were under the vigilance of some unsleeping eye.

In these descriptions, Dylan is exercising his surreal sense of humor in an absurdist listing of thematically connected objects. This delight in lists became a trademark feature of Dylan's brilliant satellite radio show, *Theme Time Radio Hour,* which ran from May 2006 to April 2009. On the show, Dylan treated listeners to an expert, thematically arranged journey through folk, blues, jazz, country, and popular song of every variety. In Episode 1.11, "Flowers," DJ Dylan treats listeners to one of these lists:

> Tonight we're going to be talking about the most beautiful things on earth, the fine-smelling, colorful, bee-tempting world of flowers, the Bougainvillea, the Passion Flower, the Butterfly Clerodendron, the Angel's Trumpets, the Firecracker plant, we're going

> to be talking about Rosa rugosa, the Angel Face, All that Jazz, the Double Delight, the Gemini [Dylan's zodiac sign] and the Julia Child, we're going to be talking about the Knockout Shrub, the New Dawn, the Mr. Lincoln—and that's only the roses—we're also going to hit on the Silver King, the German Statis, the Globe Thistle and the Joe Pie Weed, the Violet, the Daisy, the lovely Chrysanthemum, the Arrow and the Tansy, we'll be hitting on the Bachelor's Button, the Coxcomb and the Lion's Ear, the Love in the Mist and the Victoria Sorghum [laughs],—I just made that one up—we're going to be talking about "Flowers," on *Theme Time Radio Hour*. (58)

THE TOOL ROOM

Back in Ray and Chloe's apartment, another room turns out to be a workshop, with "all kinds of paraphernalia piled up" (58):

> There were some iron flowers on a spiral vine painted white leaning in the corner. All kind of tools laying around. Hammers, hacksaws, screwdrivers, electricians' pliers, wire cutters and levers, claw chisels, boxes with gear wheels—everything glistening in the backlight of the sun. Soldering equipment and sketch pads, paint tubes and gauges, electric drill—cans of stuff that could make things either waterproof or fireproof.

Here is yet another list, creating an image of a room overflowing with metalworking tools and gadgets, Dylan delighting in the detail. Dylan is knowledgeable about such metalworking matters. I suggest we're no longer in the tool room of Ray's Tribeca apartment, but rather in a version of Dylan's art studio in Los Angeles that has merged with whatever the reality was in Ray's workroom in 1961. On November 17, 2013, an exhibit of Dylan's metal sculpture was put up in the Halcyon Gallery in London and fans discovered that he had for many years been soldering and welding scrap metal objects—car parts, lawn mowers, chains, iron wheels, and so on—into artwork, particularly ornamental gates.

Dylan picked up these skills long before 2013. He has an uncredited cameo in the 1990 movie *Catchfire,* in which he has a brief encounter with hit man Milo, played by Dennis Hopper, who is looking for Jodie Foster, in the role of electronic artist Anne Benton. She has witnessed Milo's killing of a rival and he is on her trail. Hopper also directed the movie, which he disowned, releasing a cable television version under the name *Backtrack*. In a scene more or less gratuitous to the plot, Dylan is in a workshop sculpting wood with a chain saw and displaying his metal sculpture. Going even further back, and in real life, Suze Rotolo, Dylan's first real girlfriend and Muse in New York, recalls Dylan's woodworking skills in her autobiography, *A Freewheelin' Life*, published in 2008. After Dylan had bought a secondhand TV for the apartment on Fourth Street, she writes,

"The TV never really worked very well, so Bob took it out of the cabinet and used the wood to build a decent coffee table and better shelves." Such things don't happen without some training.

At that point, Bob Dylan was only two years out of high school, and he presumably picked up these skills in Hibbing, where carpentry and metalworking were of immense importance for employment in the mines, but were also available as electives. On December 14, 1956, the middle of Dylan's sophomore year, the school paper, the *Hibbing Hi Times,* contained an article "Metal Arts Class Trains Students in Use of Machine, Hand Tools," with detail that shows where it all likely started:

> Students thus far have produced a wide variety of machine and hand tools, including belt sanders, vises of various kinds, smooth planes for wood working, and a drill press. They have also made repair parts for lawn mowers and tractors, each student grinding his own bits or cutting tools.

If, as seems likely, Ray Gooch's tool room is really an allusion to Dylan's own sculptural activities, it is worth adding a further detail from the imaginative mind of Bob Dylan. In episode 21 of *Theme Time Radio Hour,* "School," aired in the fall of 2006, the year *Modern Times* came out with the lines from Roman poet Ovid, DJ Dylan seems to be back in Hibbing: "there are many different kinds of teachers," he says, naming only two, "there are Latin teachers, shop teachers." We know the name of his Latin teacher; perhaps he also picked up some skills in metal arts or the woodworking shop.

As Dylan wrote, "Ray was as mysterious as hell." So is Bob Dylan, and one senses a "transfiguration" here, a term we'll return to soon. Dylan and Ray both work with metal. Ray's girlfriend Chloe was working at the Egyptian Gardens, whereas Sara Dylan, before she met Dylan, had worked as a model and Playboy Bunny at the New York Playboy Club. And before that, Dylan's first-known New York girlfriend, Avril, whose apartment he shared in 1961, was also a dancer. Talk about parallel lives. In *Chronicles,* the stream of consciousness continues into the next room.

THE LIBRARY

Ray Gooch's library is the room in "The Lost Land" that Dylan gives the most attention. Before entering the library in this chapter, he explains his own place in the history of songwriting (34–35):

> Songs were my preceptor and guide into some altered
> consciousness of reality, some different republic,
> some liberated republic. . . . I didn't know what age
> of history we were in nor what the truth of it was.
> Nobody bothered with that. If you told the truth that

> was all well and good and if you told the un-truth, well that's still well and good. Folk songs taught me that.

This is an important moment in the book, where Dylan admits that the boundary between truth and untruth in his mind, and in his art, is indistinct. The fact that Dylan provides this signpost right before entering the library is clear indication that his creative imagination was at the wheel, just as much as his actual memory of the books he may or may not have seen in that room. Dylan describes finding himself in this library "looking for the part of my education that I never got." And a little later (35–36):

> The place had an overpowering presence of literature, and you couldn't help but lose your passion for dumbness. Up until this time I'd been raised in a cultural spectrum that had left my mind black with soot.

We know that this is an exaggeration, and that Dylan's cultural mind was hardly "black with soot" when he arrived in New York in early 1961, at the age of nineteen. We know that he'd taken B. J. Rolfzen's poetry classes at Hibbing High, and it is interesting that he claims to have "read [in Ray's library] the poetry books, mostly. Byron and Shelley and Longfellow and Poe." But of other books that he here comes across, Dylan says he's only browsed through them, rather than read them: "I would have had to have been in a rest home or something in order to do that."

What were the actual titles in that library, whether real or imagined? At the top of Dylan's list, receiving three mentions in two pages, is the ancient Greek writer Thucydides's *History of the Peloponnesian War,* which Dylan refers to as *The Athenian General*. He gets the title wrong, but no matter, for he captures the relevance of the Greek historian (36):

> It was written four hundred years before Christ and it talks about how human nature is always the enemy of anything superior. Thucydides talks about how words in his time have changed from their ordinary meaning, how actions and opinions can be changed in the blink of an eye. It's like nothing has changed from his time to mine.

Clearly Dylan has dipped into Thucydides, as we can see from similarities between his description above and Rex Warner's Penguin translation of one of the most famous passages of *History of the Peloponnesian War* (1.22):

> It will be enough for me, however, if these words of mine are judged useful by those who want to

> understand clearly the events which happened in the past and which (human nature being what it is) will, at some time or other and in much the same ways, be repeated in the future. My work is not a piece of writing designed to meet the taste of an immediate public, but was done to last for ever.

Thucydides was very much in the air in 2003 and 2004, when Dylan was writing *Chronicles* and the United States was fighting wars in Afghanistan and Iraq. Reading Thucydides then would indeed "give you the chills," as it did for Dylan. The Greek historian had said of the unwise decision of the Athenians to invade Sicily in 415 BC: "The result of this excessive enthusiasm of the majority was that the few who were opposed to the expedition were afraid of being thought unpatriotic if they voted against it, and therefore kept quiet." Dylan, writing at the time of our own wars, seems to have been thinking precisely of passages such as this, and about the relevance of the history of ancient Greece to modern America, for more than a decade, as in a 1991 interview, the second year of the First Gulf War:

> A college professor told me that if you read about Greece in the history books, you'll know all about America. Nothing that happens will puzzle you ever again. You read the history of Ancient Greece and

> when the Romans came in, and nothing will ever bother you about America again. You'll see what America is.

History is always about the place of the past in the present time, and in 1991 and 2004, the time of Gulf Wars I and II, Dylan was connecting America and the ancient Greeks and Romans. He does so in these pages of *Chronicles* that do not name any book but give an example of how the Greeks dealt with occupation in the same area where the United States currently found itself:

> Alexander the Great's march into Persia. When he conquered Persia, in order to keep it conquered, he had all of his men marry local women. After that he never had any trouble with the population, no uprisings or anything.

It is hard not to take this as surreal advice emerging from the surreal world that is "The Lost Land."

Thucydides's contemporary Sophocles, the writer of tragedies, is also there in the library, again with a wrong title, as is Tacitus, the greatest of the Roman historians, though he wrote histories, not "lectures and letters to Brutus." "The Twelve Caesars" of Suetonius, the other Roman historian on whom Robert Graves based his novel *I, Claudius*, is also there. Ray's library also contained the Roman poet Ovid's *Metamorphoses,* "the scary horror tale"—not a bad description of a work that depicts how human bodies are transformed into trees, birds, flowers, and various kinds of beasts.

Beyond the Greeks and Romans, Dylan expands his range of literary references. From the thirteenth century he mentions Dante's *Inferno,* from the sixteenth and seventeenth centuries Machiavelli's *The Prince,* Fox's *Book of Martyrs,* and Milton's poem "On the Late Massacre in Piedmont." He moves on to the nineteenth and twentieth centuries with "Gogol and Balzac, Maupassant, Hugo and Dickens," books on Mormon prophet Joseph Smith, on Confederate general Robert E. Lee, and on Sigmund Freud, "the king of the subconscious." But nothing he mentions is in chronological order. Everything is jumbled up, Ovid next to the "autobiography of Davy Crockett," Rousseau's *Social Contract* next to *Temptation of St. Anthony*. This mixing up helps lend the whole catalog an air of pure stream of consciousness, with a delectable juxtaposing of these daunting and frequently off-putting titles in the lively, contemporary voice of Bob Dylan, resulting in an incongruity essential to his humor. Thucydides "could give you the chills," "*Materia medica* (the causes and cures for diseases)—that was a good one," "The words of 'La Vita Solitaria' by Leopardi [nineteenth-century Italian poet] seemed to come out of the trunk of a tree, hopeless, uncrushable sentiments." Joseph Smith "pales in comparison to Thucydides." "Albertus Magnus was lightweight next to Thucydides," with a play or joke on the meaning of Latin *magnus* ("big"): "a lot of these books were too big to read, like giant shoes for large-footed people." "In the end," Dylan writes, "the books make the room vibrate in a nauseating and forceful way." Ray Gooch's library reflects the creative essence of Dylan's mind, unfettered by catalogs or by order, and getting to the heart of who he is artistically.

Still in the library, on page 45 of *Chronicles* Dylan writes, "Invoking the poetic muses was something I didn't know about yet." He may not have known about that in 1961, but by 2004 things had changed, as he had been reading and drawing from the classical texts. The most famous encounter between a poet and the Muses is in the Greek poet Hesiod, from the eighth century BC, whom Dylan was quoting in the Rome press conference in 2001 when he discussed the Iron Age and the Golden Age of Homer. There he also mentioned the memoir he was working on. The Muses tell Hesiod what he should learn from them: "We know how to speak many false things that seem like the truth, but we also know, when we choose, how to sing the truth." Like Dylan, they knew the truth and untruth, and both are fine.

That is what was going on in the description of Ray Gooch's library, and it is also what is going on in Dylan's songwriting, right up to the epic 14-minute, 45-verse song "Tempest," from the 2012 album of the same name. This song is really the culmination of his songwriting. There is truth in it, indeed in its opening words of the second verse, "'Twas the fourteenth day of April," the day in 1912, one hundred years before the album came out, that the *Titanic* hit the iceberg that sank it early the next morning. That's a truth, as is the fact that John Jacob Astor IV, wealthiest of the passengers, went down with the ship. "The rich man, Mr. Astor, kissed his darling wife," goes the verse. Everyone else in the song is made up. Wellington, who "strapped on both his pistols"; Calvin, Blake, and Wilson, who are "gambling in the dark"; Jim Dandy, who gave up his seat to the "little crippled child"—on the album it was Jim Backus, the actor who played Thurston Howell III on the 1960s TV series of a different shipwreck story, *Gilligan's Island*, another lost land—"Davey the brothel keeper," all of these are untruths. And that too is all well and good, all part of the song that becomes in our memory this newly empowered version of the sinking of the *Titanic*. Go back to the folk song "The Titanic," by the Carter Family, from which Dylan took the melody and most of whose words he repurposed as some of the folk song components of his fictional epic. That folk song is visible and audible, and there is no effort or intention to hide the fact. On the contrary, Dylan's song is the richer for our hearing the old song in his new song. But the new song is something else, something that through Dylan's genius as a songwriter, singer, verbal painter, has transcended the folk tradition in which it is rooted; it has become both epic and cinematic, a wholly new genre.

BACK TO REALITY

The last chapter of the book takes us back to the first and second. "River of Ice" covers much of the time period of "The Lost Land," but without the surreal essence of the earlier chapter. In fact, for an understanding of what can be known of Dylan's life, what it was like growing up in Hibbing, the move to the coffeehouses of Dinkytown, and the eventual move east, you could do worse than start with this last chapter, which ends where the first chapter began, closing the circle, John Hammond signing him to a record deal with Columbia Records in 1961 and Dylan recording the first album in Lou Levy's studio. "In my beginning is my end," as T. S. Eliot put it.

So we might expect to run into Ray and Chloe in less surreal guise in this last chapter, which treats 1961 and looks more like truth than untruth. According to *Chronicles,* Dylan met them through folksinger and folklorist Paul Clayton, whom Dylan describes in Chapter 1 as "good natured, forlorn and melancholic" (26)—with no mention of the fact that he would take his own life in 1967. Clayton himself returns in Chapter 5, and still with no mention of his fate (260–61):

> He knew hundreds of songs and must have had a photographic memory. Clayton was unique—elegiac, very princely—part Yankee gentleman and part Southern rakish dandy. He dressed in black from head to foot and would quote Shakespeare. Clayton traveled

> regularly from Virginia to New York, and we got to be friends. His companions were out-of-towners and like him, a "caste apart"—had attitudes, but known only to themselves—a non-folky crowd.

That's all we see of Ray and Chloe. They have disappeared, or rather stepped off the stage, here present only by implication as two of the "out-of-towners" with whom Dylan spent the early weeks of his New York period in early 1961. They and their apartment belonged in the lost land, and that is where Dylan leaves them in the drama of this book. The library of that land, the windowless room "with a painted door—a dark cavern with a floor-to-ceiling library," has vanished back into the mind of Dylan. Or, as he sang at the end of the melancholic "Forgetful Heart" in 2008, "The door has closed forevermore / If indeed there ever was a door."

THE TRANSFIGURATION OF BOB DYLAN

Bob Dylan's interviews and press conferences are a genre worthy of study in itself, fifty-five years' worth of creative control and orchestration of the image and information he has permitted the world to possess. He chooses the time, the place, and the interviewers. It is possession of details of his life and lyrics that Dylan's fans have craved. He almost never discusses or responds to questions about the meaning of lyrics, his politics, relationships, or, since 1980, religious affiliation. Likewise, it is dangerous to trust too much what he does let the world see. My interest in his interviews has to do with the artistic changes through which he went mostly in the twenty-first century, connecting himself across time to other artists, going right back to the ancients. A good part of him is now living in this world.

This process started early, as early as January 31, 1959, when the eighteen-year-old saw Buddy Holly play at the Duluth National Guard Armory, three days before Holly's death in a plane crash. In his Nobel lecture, delivered on June 5, 2017, Dylan tells what happened:

> He was powerful and electrifying and had a commanding presence. I was only six feet away. He was mesmerizing. I watched his face, his hands, the way he tapped his foot, his big black glasses, the eyes behind the glasses, the way he held his guitar, the way he stood, his neat suit. Everything about him. He looked older than twenty-two. Something about him seemed permanent, and he filled me with conviction. Then, out of the blue, the most uncanny thing happened. He looked me right straight dead in the eye, and he transmitted something. Something I didn't know what. And it gave me the chills.

In an interview in 1978, the year before Dylan converted to Christianity for a year or so, Jonathan Cott brings up the French

Symbolist poet Arthur Rimbaud, long associated with Dylan's music of the mid-sixties, and subject of my next chapter: "I've always associated you with Rimbaud . . . do you believe in reincarnation?" Dylan wanders through various possibilities for reincarnation, concluding: "I think one can be conscious of various vibrations in the universe. But reincarnation from the twelfth to the twentieth century—I say it's impossible." Cott modifies his question: "when I say Rimbaud and you, you take it as an affinity." Dylan: "Maybe my spirit passed through the same places as his did. We're all wind and dust anyway and we could have passed through many barriers at different times."

More recently, in December 2001, following the release of *"Love and Theft,"* with those lines of Virgil, Dylan gave another interview, with American writer and music journalist Mikal Gilmore. As the interview was winding down, Gilmore asked Dylan where the songs on *"Love and Theft"* came from, noting that the album feels like it's from "the America we live in now, but also the America we have left behind." Dylan's response was complicated:

> I mean, you're talking to a person that feels like he's walking around in the ruins of Pompeii all the time. It's always been that way, for one reason or another. I deal with all the old stereotypes. The language and the identity is the one I know only so well, and I'm not about to go on and keep doing this—comparing my

> new work to my old work. It creates a kind of Achilles heel for myself. It isn't going to happen.

Pompeii and Achilles, the world of Rome and of Homer, are mentioned as if Dylan is inhabiting the ancient places. Gilmore doesn't pick up on Dylan's references, as regularly happens with his interviews. Instead, he brings things back home to America, asking whether the album emanates from Dylan's experience of America at that moment. "Every one of the records I've made," Dylan replies, "has emanated from the entire panorama of what America is to me." That panorama included the Rome of Dylan's youth, in Latin classes, the Latin Club, and at the movies, possibly including the director Sergio Leone's 1959 movie, *Last Days of Pompeii*, the town destroyed in AD 79 by the eruption of Mount Vesuvius, which Dylan had just mentioned. In the summer of that year, 2001, Dylan had given the Rome press conference, on July 23. We'll never know, but I suspect he visited the impressive site of Pompeii, perhaps four days later, July 27, the day after performing a few miles across the Bay of Naples, and before heading off for a concert the following day in Taormina, Sicily.

Dylan's thirty-second studio album, *Modern Times,* was released on August 29, 2006, and was soon hailed as a continuation of the comeback that had begun with *Time Out of Mind* (1997) and continued with *"Love and Theft"* (2001), the last of the "trilogy," as it seemed, and was prematurely labeled by some critics. A week after the release of *Modern Times, Rolling Stone* published "The Genius and Modern Times of Bob Dylan," written by novelist Jonathan Lethem, who had been interviewing Dylan about the new album. Well before we learned about the classical and other texts in these songs, borrowings from Roman exile Ovid and confederate poet Henry Timrod, Dylan was laying down more clues and hints about the transformations, reincarnations, and transfigurations that his art was undergoing. Here is how Lethem portrayed it, starting with his quote of Dylan from the interview:

> "I just let the lyrics go, and when I was singing them, they seemed to have an ancient presence." Dylan seems to feel he dwells in a body haunted like a house by his bardlike musical precursors. "Those songs are just in my genes, and I couldn't stop them comin' out. In a reincarnative kind of way, maybe. The songs have got some kind of a pedigree to them. But that pedigree stuff, that only works so far. You can go back to the ten-hundreds, and people only had one name. Nobody's gonna tell you they're going to go back further than when people had one name."

Who knows when the ten hundreds were? Maybe the Middle Ages, maybe even further back. To those pedigreed people with one name who have been inhabiting Dylan's song in his renaissance of the last twenty years, and for even longer without his fully realizing it: Homer, Virgil, Ovid, Plutarch, Petrarch, Dante, whom Dylan claimed to know by way of Gooch's library, or more likely through his own serendipitous reading.

Then there is perhaps the liveliest interview he has ever given, in 2012 to Mikal Gilmore in *Rolling Stone* following release of the new, and to date last, original album, the masterpiece *Tempest*. Dylan talks about his own "transfiguration" and produces a book he had brought with him to the interview. He hands it over to Gilmore, who reads some of the book into his tape recorder. It is Ralph "Sonny" Barger's bestseller *Hell's Angel: The Life and Times of Sonny Barger and the Hell's Angels Motorcycle Club,* cowritten by Barger with Keith and Kent Zimmerman, or "Zimmermen," as they call themselves in the preface. The pages Dylan has Gilmore read narrate the motorcycle death in 1961 or 1964 of someone called Bobby Zimmerman. *Hell's Angel* is a true story, but for Gilmore things were getting a little strange. None of these three Zimmermans is related, at least not in the conventional sense of the word, to the Bob Zimmerman who became Bob Dylan, who connects the incident in the book to his own motorcycle accident in Woodstock (which happened in 1966, two to five years after that of Bob Z. the Hell's Angel)—though in a suggestive rather than specific way.

He is explaining his own changes in general. To the question, "Are you saying that you really can't be known?" Dylan replies:

> Nobody knows nothing. Who knows who's been transfigured and who has not? Who knows? Maybe Aristotle? Maybe he was transfigured? I can't say. Maybe Julius Caesar was transfigured. I have no idea. Maybe Shakespeare. Maybe Dante. Maybe Napoleon. Maybe Churchill. You just never know because it doesn't figure into the history books. That's all I'm saying.

Julius Caesar, Aristotle, and Dante—again we are back in the world of the Greeks and Romans and their greatest late medieval inheritor, Dante, Italian poet of the thirteenth century. The notion of transfiguration is never quite explained, can't really be explained. When pushed by Gilmore, Dylan responds as usual: "I only know what I told you. You'll have to go and do the work yourself to find out what it's about." And to do that you would have to find "a book about transfiguration." There is no such book. We are in the world of untruth, as Dylan in this strange and strangely enjoyable interview heads back to—where else?—Rome, the place of his original transfiguration, the city "where I was born" ("Going Back to Rome"). He tells Gilmore:

> About a year later I went to a library in Rome and I found a book about transfiguration, because it's nothing you really hear about every day, and it's in the mystical realm, and I found out only enough to know that, uh, OK, I'm not an authority on it, but it kind of sets you straight on what sets you apart.

The focus of the interview at one point turns from *Tempest* to the quality and staying power of his last five albums. Everything since *Time Out of Mind,* notes Gilmore, "is a body of work that can stand on its own." Dylan uses this as an opportunity to talk about his music in a way that is at the heart of what this book is about:

> The thing about it is that there is the old and the new, and you have to connect with them both. The old goes out and the new comes in, but there is no sharp borderline. The old is still happening while the new enters the scene, sometimes unnoticed. The new is overlapping at the same time the old is weakening its hold. It goes on and on like that. Forever through the centuries.

There are different ways of interpreting this, and Dylan goes on to talk about the shifts in his work from the 1950s to the 1960s, but the words "forever through the centuries" are pretty explicit, and in sync with what is happening with his songs. Later in the interview he will talk in the same way about his performance practices: "[i]t's always been this way for everybody who's ever done it, going back to those ancient days."

In 2017, when Dylan was about to release his thirty-eighth studio album *Triplicate,* he did another of the carefully scheduled interviews that occur on such occasions, with Bill Flanagan, who at one point says: "No one can hear 'As Time Goes By' and not think of *Casablanca.* What are some movies that have inspired your own songs?" Like the songs he sings on these albums, Dylan's response takes us back to his teenage years in Hibbing and the world of Roman centurions, gladiators, and biblical epics: "*The Robe, King of Kings, Samson and Delilah.*" He could have seen *The Robe* at the State Theater in Hibbing in January 1954. The end of the interview moves from the Roman to the Greek world, as Flanagan asks whether the title *Triplicate* brings to mind Frank Sinatra's trilogy of 1980, *Past Present Future*. "Yeah, in some ways, the idea of it," Dylan replies, adding, "I was thinking in triads anyway, like Aeschylus, *The Oresteia,* the three linked Greeks plays. I envisioned something like that." A follow-up would have been interesting, but the interview instead moved on. We will return to triads.

BOB DYLAN, ROMAN HISTORY TEACHER

Following on from the readings in Ray Gooch's library, in a 2009 interview with historian Douglas Brinkley, Dylan is asked about the importance of Christian scripture in his life. He redirects the discussion to more works from the Greek and Roman canon:

> [T]hose other first books I read were really biblical stuff. *Uncle Tom's Cabin* and *Ben-Hur.* Those were the books that I remembered reading and finding religion in. Later on, I started reading over and over again Plutarch and his *Roman Lives.* And the writers Cicero, Tacitus, and Marcus Aurelius . . . I like the morality thing. People talk about it all the time. Some say you can't legislate morality. Well, maybe not. But morality has gotten kind of a bad rap. In Roman thought, morality is broken down into basically four things. Wisdom, Justice, Moderation and Courage. All of these are the elements that would make up the depth of a person's morality. And then that would dictate the types of behavior patterns you'd use to respond in any given situation. I don't look at morality as a religious thing.

I suspect Dylan picked Brinkley as an interviewer—his only interview with a full-time academic—because it was important that his words on this topic, which are fully coherent and have the ring of truth and sincerity, not be garbled or misread.

In 2015, Dylan did a very smart and musically engaged interview with the editor in chief of *AARP The Magazine,* Robert Love. The interview itself appeared in the March/April issue. To find the theme that has been running through the other recent interviews, to get back to Rome, you have to do some digging and locate the version of the interview on AARP's website, along with its accompanying photo slide show. In the last of a gallery of twelve slides revealing themes from the interview and its primary subject is a photo with the following caption:

Bob Dylan: His True Calling
"If I had to do it all over again, I'd be a schoolteacher—probably teach Roman history or theology."

That sounds about right. As for the slide itself, all it shows is an open book, resting on a stack of four other books. Three of the four dog-eared volumes in the stack look old, going back to the nineteenth or early twentieth century, the sort of books a teacher of Roman history or theology would use, also the sort of books you might have found in the library of Ray Gooch, old editions of Cicero, Tacitus, and Plutarch. The website is that of the AARP, but the hand of Bob Dylan is at work here.

5

THE EARLY THEFTS: "MINE'VE BEEN LIKE VERLAINE'S AND RIMBAUD'S"

SOME STUFF I'VE WRITTEN, SOME STUFF I'VE DISCOVERED, SOME STUFF I STOLE.

—DYLAN TO JOHN HAMMOND

Allusion, reference, plagiarism—these are all names for the phenomenon known as "intertextuality," a term that is most convenient in its neutrality for describing the process by which poets, songwriters, painters, composers, or artists of any genre produce new meaning through the creative reuse of existing texts, images, or sound. In its truest sense, intertextuality is as far as you can get from plagiarism, which is a practice meant to escape notice. Plagiarism is about passing off as your own what belongs to others. In contrast, the most powerful and evocative instances of intertextuality enrich a work precisely because, when the reader or listener notices the layered text and recognizes what the artist is reusing, that recognition activates the context of the stolen object, thereby deepening meaning in the new text.

This is a very old phenomenon, as Dylan came to realize. In Homer's *Odyssey,* the Greek hero goes down to the Underworld to consult the ghost of the seer Tiresias about how he is to get back home to Ithaca. There Odysseus also meets the ghost of his mother. He talks with her, then tries to embrace her:

> Three times I rushed toward her, desperate to hold her,
> Three times she fluttered through my fingers, sifting
>   away
> Like a shadow, dissolving like a dream.
>
> —*Odyssey* 11.235–37, tr. Fagles

Seven hundred years later, in the *Aeneid,* published in 19 BC, the Roman poet Virgil sends his hero Aeneas down to the underworld, where he meets the shade of his father, at which point he tries and fails to carry out a similar embrace, with a similar result:

> Three times he tried to fling his arms around his neck,
> Three times he embraced—nothing, the phantom
> Sifting through his fingers,
> Light as wind, quick as a dream in flight
>
> —Virgil, *Aeneid* 6.808–11, tr. Fagles

Virgil, accused of plagiarizing Homer in his own day, has in a sense done so, though from one language to another, and he means us to recognize the intertextuality, to see the loss of Greek Odysseus in the loss of Roman Aeneas, two heroes grieving for their parents, a universal scene of shared humanity. Additionally, Virgil had already put the lines to work at the end of his *Aeneid* 2, where he used them to describe Aeneas's attempt to embrace the ghost of his wife Creusa, lost in the flames of Troy. Virgil *wants* the reader to notice his borrowings, and to be enriched by those added intertexts: the loss of a mother, a father, a wife, each is a new version of the others, expressed in the same lines to convey the same grief. This is how literature works, and it is how Dylan's song works when it takes on the songs and texts that are in his tradition.

Much has been made of Dylan's "borrowings," from early on, and how you look on them in part depends on how you think literature and art in general should work, particularly on whether you insist on notions of "originality," as if anything rooted in folk, blues, and poetry at large is ever wholly original. Dylan is in a tradition that is old and his connecting to those traditions is a big part of what his art is about.

Let's look at a couple of early cases, one with Dylan stealing, the other with Dylan being stolen from, and look at the difference. In 1963, for the song "Masters of War," Dylan took the striking melody of folksinger Jean Ritchie's "Nottamun Town," a song that Ritchie herself considered one of her "family songs," a proprietary artifact. It had been collected by the English folk song collector and folk dance promoter Cecil Sharp in 1917. His primary source? Jean Ritchie's great-aunt Una. Apparently brought to America by her great-great-grandfather Crockett, the song was duly recorded by Jean on her self-titled 1960 album. Ritchie was none too pleased when Dylan borrowed the melody for his song, a song whose lyrics bore no resemblance to the words of the song that Ritchie saw as her property. The melody had been around for decades in Appalachia, and for centuries back in the English Midlands. "Nottamun" is most likely a dialect version of "Nottingham," and the song perhaps goes back to the seventeenth century and the time of the English civil wars. Folk song does not belong to anyone, and even though Ritchie claimed Dylan settled the case out of court, Dylan did not acknowledge the justice of her case. By any standard, it is a good thing that Dylan heard that tune, which was filtered through his own genius and gave us the song "Masters of War."

Both songs will ultimately survive and stand the test of time. "Nottamun Town," an ancient and eerie song that goes back to the Middle Ages, is beautiful and archaic sounding, as sung not just by Ritchie, but also for instance by the English folk group Fairport Convention. But "Masters of War" is a song that has been around for the past fifty-five years and still feels timely and immediate in its relevance, and will feel so for as long as older men profit from the wars in which young men die.

Dylan saw that the melody he stole was right for this song, so he stole it. And can the melody of a folk song like "Nottamun Town" really belong to anyone? Yet modern copyright laws do exist to avoid disputes like this one.

The tables would turn in 1995, when, according to the cable television network VH1, Dylan was said to have settled out of court with Darius Rucker of Hootie & the Blowfish over the use of lyrics from "Tangled Up in Blue" in Darius's song "Only Wanna Be with You":

Put on a little Dylan sitting on a fence
I say that line is great, you ask me what I meant by
Said, I shot a man named Gray, took his wife to Italy
She inherited a million bucks and when she died it
came to me
I can't help it if I'm lucky
I only wanna be with you
Ain't Bobby so cool
I only wanna be with you
Yeah I'm tangled up and blue
I only wanna be with you

Artistically, there is a world of difference between this instance of direct quotation, the italicized words, from "Tangled Up in Blue," and Dylan's borrowing in "Masters of War." In Rucker's song, Dylan's words have no new life of their own, and no artistic function other than as direct, unauthorized quotation in a context that was purely commercial in nature.

Dylan has been stealing since the very beginning. On May 1, 1960, three weeks before his nineteenth birthday, at the time a freshman at the University of Minnesota, Dylan performed twenty-seven songs in the home of Karen Wallace, in St. Paul. This was at the start of his folksinging phase, and none of these songs was a Dylan original; they were just folk songs. Dylan had recently read Woody Guthrie's *Bound for Glory* and he was soaking up Guthrie, Leadbelly, Pete Seeger, and other folksingers. He had stopped going to class, in favor of spending time listening, talking, performing at the Ten O'Clock Scholar, a coffeehouse in bohemian Dinkytown, a few blocks from the university. The songs Dylan sang for Wallace were mostly traditional pieces—cowboy songs, traveling songs, gambling songs, girls-left-behind songs, prison songs, gospel, and blues. Some were Guthrie originals, notably "This Land Is Your Land" and "Pastures of Plenty," and others had been recorded by Guthrie or other folksingers the young Dylan was assimilating. There is nothing remarkable about his singing of those songs. He was "covering" them as singers have always done with traditional material. But as his art developed, that would all change. They would provide the elements of his original songwriting, their traces visible but transformed in the process of his own songwriting.

One of those twenty-seven songs, "Columbus Stockade," goes back to the 1930s and beyond. Dylan had surely heard the

Hank Williams or the Guthrie version, which eventually came out on Guthrie's 1964 album, *The Early Years*. In the song, the narrator is in jail in Columbus, Georgia, and he seems to have been let down, betrayed by friends and by a lover who's gone off with another man:

Way down in Columbus Stockade
Oh to be back in Tennessee
Way down in Columbus Stockade
Where my friends went back on me

You can go and leave me if you want to
Never let me cross your mind
In your heart you love another
Leave me darling, I don't mind

Way down in Columbus Stockade
Left me there to lose my mind
Thinking about my blue-eyed honey
Purtiest girl that I left behind

You can go and leave me if you want to
Never let me cross your mind
In your heart you love another
Leave me darling, I don't mind

The song stayed in Dylan's repertoire after this first performance in 1960, and he would play it when he arrived in New York the next year. As he explains in *Chronicles: Volume One* (18):

> Folk songs were the way I explored the universe, they were pictures and the pictures were worth more than anything I could say. I knew the inner substance of the thing. I could easily connect the pieces. It meant nothing for me to rattle off things like "Columbus Stockade," "Pastures of Plenty," "Brother in Korea," and "If I Lose, Let Me Lose" all back-to-back like it was one long song.

Dylan's first live-in (in her apartment) girlfriend in New York was a dancer from California named Avril. In the spring of 1961, we know that he left her alone in New York to head back home to Minnesota for a month. She had apparently told him that she too would be making a trip home to California while he was gone, but he seems to have been surprised and upset when he returned to find her gone, and he wrote her a song in response. "California Brown-Eyed Baby" was one of his earliest original songs, and he sang it to her over the phone, an emotional event as recalled by Eve MacKenzie, in whose Greenwich Village apartment Dylan was living off and on in 1961. Dylan's song is written to the tune of "Columbus Stockade":

The rain is fallin' at my window
My thoughts are sad forever more
I'm thinkin' about my brown-eyed darlin'
The only one that I adore

She's my California brown-eyed baby
The only one I think about today
She's my California brown-eyed baby
Livin' down San Francisco way

Sadly I look out of my window
Where I can hear the raindrops fall
My heart is many thousand miles away
Where I can hear my true love call

Now boys don't you start to ramble
Stay right there in your home town
Find you a gal that really loves you
Stay right there and settle down.

This song is more than a simple ploy to get his girl back. It is an early example of the art of his songwriting, traditional but also original. Dylan's narrator is "thinkin' about my brown-eyed darlin'" as the jailed narrator in Columbus Stockade is "thinking about my blue-eyed honey." Those words and their rhythm, along with the tune, are the main markers of the theft and the sign of intertextuality, the invitation to examine the intertext, in this case "Columbus Stockade," and see if any of its lyrics are activated by the new version or elsewhere in Dylan's early songbook.

"Columbus Stockade" finds much of its power and interest in the bitterness of its repeated refrain: "In your heart you love another / Leave me darling, I don't mind." The absence of any such bitterness from "California Brown-Eyed Baby" suggests a more casual attitude to Avril, whose main appeal may have had to do with her apartment, a welcome upgrade from the couches and spare rooms he found to lay his weary head in New York City in 1961. But things would change when he met Suze Rotolo, Muse and real girlfriend, later in the year. The song that he wrote the next year when *she* left him, to study in Italy, was one of the greatest he ever wrote, "Don't Think Twice, It's All Right." It had a different tune, but the last line from "Columbus Stockade," held back from the song to Avril—"Leave me darling, I don't mind"—turns up first in the title of "Don't Think Twice, It's All Right": in each we find an order followed by a reason, comma in between, three words on either side of it. If Dylan alludes to and therefore invokes the context of the folk song with his title, he actually comes out and quotes words and sentiment that he had not bothered about with Avril's song: "You could have done better but *I don't mind.*"

"Columbus Stockade"	"Don't Think Twice, It's All Right"
You can go and leave me if you want to Never let me cross your mind In your heart you love another *Leave me darling,* **I don't mind**	I ain't sayin' you treated me unkind You could have done better but **I don't mind** You just kind wasted my precious time *But don't think twice, it's all right*

In "Don't Think Twice, It's All Right" Dylan shows his genius and mastery of songwriting, but he also shows that what is understood by intertextuality is at the heart of it. The words and the thought those words express come from a continuum of texts and songs, connected by melody, lyrics, or even word arrangement of song title.

The same may be said of the intertexts in "Bob Dylan's Dream," a song that Dylan wrote in 1963, before his twenty-second birthday, about nostalgia and yearning for the friends of one's youth, long scattered to the winds:

While riding on a train goin' west
I fell asleep for to take my rest
I dreamed a dream that made me sad
Concerning myself and the first few friends I had

If the thought and sentiment of this song seem too old for the young man writing it, who regularly returned to Hibbing in the days when he was still seeing some of those first few friends, that is because it is borrowed from a traditional ballad, "Lady Franklin's Lament." Dylan has acknowledged that he got the tune and much of the lyrics from the English folksinger Martin Carthy, whom he heard singing it when he was in London at the end of 1962. The song, which goes back to 1850, is about the disappearance of Sir John Franklin, looking for the Northwest Passage in Baffin Bay in 1845:

We were homeward bound one night on the deep
Swinging in my hammock I fell asleep
I dreamed a dream and I thought it true
concerning Franklin and his gallant crew

The twenty-one-year-old, in a dark December in London, expresses a longing for the friends of his youth, but he does so by evoking the words of that old song:

How many a year has past and gone
And many a gamble has been past and won
And many a road taken by many a friend
And each one I've never seen again

The words of Lady Franklin in the mid-nineteenth-century folk song, "Ten thousand pounds would I freely give / To know on earth that my Franklin lives," become in Dylan's rewriting "Ten thousand dollars at the drop of a hat / I'd give it all gladly if our lives could be like that." Not much money to get back those lost years; by keeping the mid-nineteenth-century amount of the old song, though dollar for pounds, Dylan is inviting exploration of the relationship to the original, pointing to his source, part of the intertextual play.

Sometimes just a few words or phrases are enough to create an intertextual connection, as in Dylan's "Kingsport Town," recorded in 1962 but not officially available for almost thirty years, when it came out in 1991 on *The Bootleg Series Volumes 1–3*. It has never been included in the official lyrics books, including the most recent edition from 2016, *The Lyrics: 1961–2012,* but the song appears on Dylan's official website with the designation "Written by Bob Dylan (arr.)." "Kingsport Town" is sung to the tune of Woody Guthrie's "Who's Going to Shoe Your Pretty Little Feet" (1938), which itself goes back a lot earlier, ultimately to Scottish folk traditions. Guthrie's song is a failed persuasion song, in which the male narrator asks four questions, in a sequence moving from innocent or practical to seductive:

Who's gonna shoe your pretty little feet?
Who's gonna glove your hand?
Who's gonna kiss your red ruby lips?
Who's gonna be your man.

To the first three questions, a female voice in turn responds:

Papa's gonna shoe my pretty little feet
Mama's gonna glove my hands
Sister's gonna kiss my red ruby lips
I don't need no man

This exchange opens and closes the song, with the center containing the lament of the man whose woman is gone:

The fastest train I ever did ride
Was a hundred coaches long,
And the only woman I ever did love
Was on that train and gone

On that train and gone, boys,
On that train and gone,
And the only woman I ever did love
On that train and gone.

In "Kingsport Town," Dylan borrows the tune, and also borrows "gloves" for the second line as a sort of "footnote" to the source in Guthrie's song, alluding along with the melody to his source. His opening verse will also be used as the closing verse, creating a form of what is called "ring composition," a very elemental poetic structure going back to the Greeks and Romans:

The winter wind is a blowing strong
My hands have got no gloves
I wish to my soul that I could see
The girl I'm a-thinking of

But Dylan utterly transforms the song. His narrative picks up where Guthrie left us: the girl is gone, the singer is on the run, "a high sheriff on [his] trail" because of unspecified problems connected to falling for a "curly-headed dark-eyed girl." The repeated "Who's gonna" questions of Guthrie's frame are in the center of "Kingsport Town," and they are very different questions, only one of them picking up on the model: "Who's a-gonna be your man?" with "Who's gonna . . . ?" itself the ultimate intertextual trace in the new song. Now, in Dylan's song, the question has nothing to do with persuasion, as in the original, but is rather replaced by his imagining the girl he fell for as being with another man:

Who's a-gonna stroke your cold black hair
And sandy colored skin
Who's a-gonna kiss your Memphis lips
When I'm out in the wind
When I'm out in the wind, babe
When I'm out in the wind
Who's a-gonna kiss your Memphis mouth
When I'm out in the wind

The answer? Not papa, mama, or sister, but someone else, maybe a fellow student of Suze Rotolo, off in Italy, studying art in Perugia, while Dylan is gloveless, out in the wind of New York City. This is an exquisite piece of intertextual writing, recorded at the same session as "Don't Think Twice, It's All Right." Both songs gain much of their dynamism from the old songs that gave Dylan the idea for the new classics his experience and imagination created out of those source texts.

COMPETITIVE INTERTEXTUALITY: CONFRONTING JOHN LENNON AND DONOVAN

Dylan's song "Fourth Time Around" (from *Blonde on Blonde,* February 1966) and its relationship to the Beatles' "Norwegian Wood" (from *Rubber Soul,* December 1965) are fairly well known but worth mentioning in the larger context of Dylan's thefts and intertexts. The song has been described as "Bob Dylan impersonating John Lennon impersonating Bob Dylan," but it is also a song in which Dylan triumphs in the battle that he wages with Lennon. When you listen to "Fourth Time Around" and then go back to "Norwegian Wood," the Beatles song sounds coy, almost innocent in comparison to the sophistication of Dylan's voice and lyrics on the classic 1966 album. It is hard to imagine Dylan actually singing this to Lennon, which he apparently did, and it is very easy to believe reports of Lennon being unhappy at what must have seemed like mockery and parody. Dylan outdoes, accentuates, overloads the rhymes, and on one level does parody the simple rhyme of the Beatles song. Here are the relevant lyrics of the two songs side by side:

The Beatles' "Norwegian Wood"	**Bob Dylan's "Fourth Time Around"**
I sat on a rug biding my **time** Drinking her **wine** We talked until two and then she **said** "It's time for **bed**" And when I awoke I was **alone** This bird had **flown** So I lit a fire Isn't it **good** Norwegian **wood**?	I stood there and **hummed** I tapped on her **drum** and asked her how **come** And she buttoned her ***boot*** And straightened her **suit** Then she said "Don't get **cute** And I tried to make sense Out of that picture of you in your wheelchair That leaned up against . . . Her Jamaican **rum** And when she did **come**, I asked her for **some** She said, "No, **dear**" I said, "Your words aren't **clear** You'd better spit out your **gum**"

If you accept the fact that Dylan's song is directed at Lennon and the Beatles, Dylan's "I never asked for your crutch / Now don't ask for mine" is devastating. He's essentially telling the Beatles, "Stay away from what I'm doing." Dylan must have felt that the Beatles, and John Lennon as the songwriter with whom he was likely the most competitive, had to be confronted.

The year before saw a preview of this side of Dylan, the artist whose aim is to stay at the peak of Parnassus. Dylan met the Scottish singer-songwriter Donovan during his 1965 tour of England, parts of which were filmed by D. A. Pennebaker for the documentary *Don't Look Back* (1967). Donovan, five years Dylan's junior, is part of a group gathered in Dylan's hotel room. At one point, Donovan starts playing guitar and launches into his song "To Sing for You." "That's a good song, man," says Dylan before Donovan even finishes, then grabs his guitar and in response delivers a version of "It's All Over Now, Baby Blue," singing "You must leave now, take what you need, you think will last. . . ." Donovan listens while nervously smoking a cigarette, and he seems to get the point, particularly apparent as Dylan looks directly at him and sings the closing line as if it were written for Donovan, a verdict on the folk traditions that Dylan's music is rendering obsolete: "And it's all over now, Baby Blue." It's not a particularly easy scene to watch—it's hard to imagine John Lennon taking Dylan's private performance of "Fourth Time Around," of which we have no record, much differently.

LOST AND FOUND IN TRANSLATION: ARTHUR RIMBAUD

After *Time Out of Mind* (1997), and especially in the songs that followed in the twenty-first century, Dylan began integrating translations of non-English texts and the worlds they evoke into his song. This was a process that had begun years before with exposure to the writings of French symbolist poet Arthur Rimbaud (1854–1891).

In *A Freewheelin' Time,* her memoir of her days in Greenwich Village before, during, and after her relationship with Dylan, Suze Rotolo recalls reuniting with Dylan and catching up after their separation, when they both returned to New York in early 1963:

> He had traveling tales to tell, opinions to express, more songs to sing, and I had found other artists, poets, and music to add to my roster of enthusiasms to share. I was reading poetry by Rimbaud and it piqued his interest.

Folksinger and native New Yorker Dave Van Ronk, one of Dylan's early contacts and minders in the Village, also recalls recommending Rimbaud to Dylan early on, and claimed to have seen a book of Symbolist poetry in Dylan's apartment, a claim not inconsistent with Rotolo's memory. Perhaps even more interesting is the recollection of photographer and musician John Cohen, who recalls Dylan showing him the words of "A Hard Rain's A-Gonna Fall" in September 1962, when Suze was off studying in Italy: "I said 'Bob, if you are going to do that kind of thing you should look at Rimbaud and Verlaine.'" Dylan is probably the most reliable source on the matter, even in *Chron-*

icles: Volume One (288): "Someplace along the line Suze had also introduced me to the poetry of French Symbolist poet Arthur Rimbaud." Memory is a creative thing, especially given what is here at stake, introducing Bob Dylan to Arthur Rimbaud.

Generally speaking, the French Symbolist poets, including Arthur Rimbaud and Paul Verlaine, were interested in describing the effects of things rather than the things themselves. They focused on the confusion of those effects, and the chaotic images that can be conjured up by that fusing and confusing in their work. Rimbaud and Verlaine are known for having a tempestuous relationship, with the two living a dissolute lifestyle that included the frequent and liberal use of hashish and alcohol. Rimbaud stopped writing at the age of twenty-one, by which time he had produced an impressive body of poetry and prose, and Verlaine eventually ended up in jail for shooting Rimbaud in the hand. The changes Dylan was going through from 1964 to 1966 included wine and marijuana, perhaps other drugs, but also include making contact with Rimbaud, whose poetry came to his attention just at the right time. Dylan had no real models in what he was doing with his lyrics and his music in these years. The Beats were now more or less a thing of the past, or at least on the way out, and it was the joy of immersion in chaotic language, in contradictory and nonlinear thought, images of sound and light, that Rimbaud seems to have shown him. There was nothing like it in the English language at the time.

In Dylan's 1974 song "You're Gonna Make Me Lonesome When You Go" he pairs the two French poets:

Situations have ended sad
Relationships have all been bad
Mine've been liked Verlaine's and Rimbaud

By 1974, Rimbaud had surely not meant much to Dylan for some years—in his draft of the song in the new Bob Dylan Archive, housed at the University of Tulsa, he misspells the name "Rimbeau," of which too much should not be made in a notebook—but that made for a great verse, and in and of itself guaranteed that biographies of Dylan would put "Rimbaud" in their indexes.

Dylan was also attracted to Rimbaud's construction of a poetic identity separate from the identity of the person—"I is an *other*"—the exhortation to separate the "I" in a poem or song from the identity of the singer. The singer is always distinct from the song, and Dylan had long known that, but it must have been reassuring to find an expression of that reality in the words of the French poet to whom he was strongly attracted. As he wrote in *Chronicles: Volume One* (288):

> I came across one of his letters called "Je est un autre," which translates into "I is someone else." When I read those words the bells went off. It made perfect sense. I wished someone would have mentioned that to me earlier.

The principle he expressed there was confirmed at a concert at New York's Philharmonic Hall on Halloween in 1964, when he addressed the audience: "I'm wearing my Bob Dylan mask"—which can be heard on *The Bootleg Series Volume 6: Bob Dylan Live 1964*. And in a television press conference Dylan gave in San Francisco on December 3, 1965, Rimbaud was still at the top of his list when he was asked about his favorite poets, even if by then he was being less than forthright in responding to questions about what he was reading or who was influencing him. Here he immediately lapses into the absurd, putting on the mask:

> What poets do you dig?
>
> *Rimbaud, I guess; W. C. Fields; the family, you know, the trapeze family in the circus; Smokey Robinson; Allen Ginsberg* [the camera moves to Ginsberg, in the audience, looking very serious]; *Charlie Rich—he's a good poet.*

Dylan had reached the point where he was not willing to give away too much about what he was reading or what was going into his songs. A few months later, on March 12, 1966, he told Robert Shelton, in an interview for the biography Shelton was writing,

> *Rimbaud? I can't read him now. Rather read what I want these days. "Kaddish"* [by Ginsberg] *is the best thing yet. Everything else is a shuck. I never dug Pound or Eliot. I dig Shakespeare.*

As for the presence of Rimbaud in Dylan's songs, in February 1964, on a visit to New Orleans, Dylan publicly embraced his influence: "Rimbaud's where it's at. That's the kind of stuff that means something. That's the kind of writing I'm gonna do." Dylan was at this point writing "Mr. Tambourine Man"— "Take me on a trip upon your magic swirlin' ship," plausibly traced to Rimbaud's poem "The Drunken Boat" ("Le Bateau Ivre"). Just these eight lines from Rimbaud's hundred-line poem are sufficient to show the appeal they would have had for Dylan:

> And afterwards down through the poem of the sea,
> A milky foam infused with stars, frantic I dive
> Down through green heavens where, descending
> pensively,
> Sometimes the pallid remnants of the drowned arrive.
>
> Where suddenly the bluish tracts dissolve, desire
> And rhythmic languors stir beneath the day's full glow.

Stronger than alcohol and vaster than your lyres,
The bitter humours of fermenting passion flow.
—"The Drunken Boat," tr. Norman Cameron, 1942

Rimbaud's vivid, dreamlike images and disarrangement of the senses, with everything all piled up to the breaking point, and the poet in the midst of a natural world beyond control, but in which he revels—this is the essence of the art that Dylan came across sometime in 1962 or 1963, and it is indeed the essence of the art of "Mr. Tambourine Man," particularly in the poetry of its joyous final stanza, capturing the dynamism of what Dylan had become:

Then take me disappearing through the smoke rings of my mind
Down the foggy ruins of time, far past the frozen leaves
The haunted, frightened trees, out to the windy beach
Far from the twisted reach of crazy sorrow
Yes, to dance beneath the diamond sky with one hand waving free
Silhouetted by the sea, circled by the circus sands
With all memory and fate driven deep beneath the waves
Let me forget about today until tomorrow.

The words are all Bob Dylan's, but Rimbaud helped show him how they could be assembled. This is a new form of intertextuality, no longer verbal or reusing specific phrasing, but more aesthetically tuned and almost spiritual. Rimbaud's way of seeing the world left its imprint on some of Dylan's best-known and most classic songs of the period, in addition to "Mr. Tambourine Man," "A Hard Rain's A-Gonna Fall," and above all, I would say, "Chimes of Freedom," first performed on that same road trip in February 1964, recorded on June 9, 1964, and released on August 8 of the same year on *Another Side of Bob Dylan*.

In "Chimes of Freedom," Dylan's narrator and a friend are caught in a thunderstorm and "duck inside the doorway" of a church. As we'll see, the location matters and the church bells and lightning bolts are fused with each other as the song unfolds in six verses, each with eight lines that proceed in closely ordered fashion: four lines describing what is going on with the two observers watching the storm, and four lines in which the tolling bells and the lightning come together in a symphony, upheld by the "chimes of freedom flashing." The writer Paul Wolfe has attacked the song for raising "bewilderment to the highest degree," but there was nothing quite like it in Dylan's work to date, and it represented a fundamental artistic shift. I quote just the opening verse. What is most striking is the vast empathy Dylan summons up for a whole array of those for whom the chimes of freedom are tolling:

Far between sundown's finish an' midnight's broken toll
We ducked inside the doorway, thunder crashing
As majestic bells of bolts struck shadows in the sounds
Seeming to be the chimes of freedom flashing
Flashing for the warriors whose strength is not to fight
Flashing for the refugees on the unarmed road of flight
An' for each an' ev'ry underdog soldier in the night
An' we gazed upon the chimes of freedom flashing

The song is evidence of how Dylan's genius was working in these years, combining something traditional with the new poetic outlook he had seen in the French Symbolist poet from the nineteenth century. Dave Van Ronk plausibly claims Dylan got the song from him. In fact he got the idea for *part* of the song from Van Ronk, whose New York City roots went back a few generations:

> Bob Dylan heard me fooling around with one of my grandmother's favorites, "The Chimes of Trinity," a sentimental ballad about Trinity Church. . . . He made me sing it for him a few times until he had the gist of it, then reworked it into "Chimes of Freedom."

"Chimes of Trinity" is a song by M. J. Fitzpatrick, from 1895, and the chorus is pretty much as Van Ronk remembers it:

Tolling for the outcast tolling for the gay
Tolling for the millionaire and friends long pass'd away
But my heart is light and gay
As I stroll down old Broadway,
And I listen to the chimes of Trinity.

This is essentially the template for lines 5–8 of each of Dylan's verses. Van Ronk only remembered the chorus, so that is all he passed on to Dylan, through an oral tradition that had stripped the song of its individual verses, just leaving a chorus for which Dylan's creative imagination would find new components. Dylan had to look elsewhere for his series of four-line stories leading into the chorus. In part these came from his own fertile imagination and frenzied poetic visions. But these lines, which do not come from "Chimes of Trinity," have also struck readers as reminiscent of Rimbaud, as described by Mike Marqusee in his book on the politics of Dylan's art:

> Each verse begins with four lines adumbrating a single conceit (at some length and often with needless convolution): the fusing of thunder, lightning, and church bells. It's a self-conscious exercise in the "disarrangement of the sense" recommended by Rimbaud and the French symbolists whom Dylan was reading at the time.

So what poem of Rimbaud's was Dylan borrowing? I propose that it's one Rimbaud wrote in 1871, at age sixteen, called "Poor People in Church." It's a blistering expression of youthful contempt, of which one translator has said, "Few adolescent rebellions have yielded such a harvest of vitriolic verse." In Scottish poet Norman Cameron's translation (the version that Dylan is likely to have read), Rimbaud goes through the different contemptible "types" in his church scene:

The timid ***ones****, the epileptic* ***one****, from whom*
Yesterday at the cross-roads people turned aside,
The blind ***ones****, nosing at old missals in the gloom,*
Who creep into the court-yards with a dog for guide

We hear some of this in Dylan's "Chimes of Freedom," where chimes toll for the "deaf an' blind." But more interesting is Dylan's style, which echoes Rimbaud's use of "one" (timid ones, epileptic one, blind one) in his fifth verse:

Electric light still struck like arrows, fired but for the
  ones
Condemned to drift or else be kept from drifting
Tolling for the searching ***ones***

And again, in the sixth:

Tolling for the aching **ones** *whose wounds cannot be nursed*
For the countless confused, accused, misused, strung-out **ones**
an' worse

"Far between" is how "Chimes of Freedom" begins, while Cameron's translation of "Poor People in Church" starts out "between," the opening verse setting the scene, as in Dylan:

Between oak benches, in mean corners stowed away,
Warming the air with fetid breath, fixing their vision
On the gilt-dripping chancels twenty mouths, which bray
The pious canticles with meaningless precision

Similarly, in verse 8 of Cameron's Rimbaud, with another anticipation of Dylan's "Far between" opening:

Far from the smells of meat, the smells of musty serge,
Prostrate and sombre farce in loathsome pantomime.
And now the worship blossoms with a keener urge
The mysticalities become still more sublime.

Most of these lines of Rimbaud are singable to the tune of "Chimes of Freedom," but only in Cameron's 1942 translation.

The difference in tone between his song and that of "Poor People in Church" is a marked one, and therein lies the difference between Dylan and Rimbaud, who had by 1966 outlived his purpose for Dylan. The disgust and contempt of the French poet for "the timid ones, the epileptic ones . . . the blind ones" that permeates Rimbaud's poem is replaced by the sheer empathy of the singer and his friend or lover in Dylan's song, "starry-eyed an' laughin'" as they listen one last time to those bells, "Tolling for the aching ones whose wounds cannot be nursed / For the countless confused, accused misused, strung-out ones and worse."

No one can really know who Bob Dylan is, but it is here worth quoting Klas Ostergren, member of the Swedish Academy present at the small gathering when Dylan picked up his Nobel Prize medal: "it went very well. He was a very nice, kind man." In this difference of outlook between the French poet and the American singer may lie the answer to why Dylan, within just a couple of years, said of Rimbaud, "I can't read him now." Rimbaud wrote no poetry after the age of twenty-one. Bob Dylan was just getting started.

6

THE GIFT WAS GIVEN BACK: *TIME OUT OF MIND* AND BEYOND

ON SOME NIGHT WHEN LIGHTNING STRIKES,
THIS GIFT WAS GIVEN BACK TO ME AND I KNEW IT. . . .
THE ESSENCE WAS BACK.

—INTERVIEW WITH ROBERT HILBURN, *LOS ANGELES TIMES*, DECEMBER 14, 1997

There is not much intertextuality in Dylan's work in the decades during and after his retirement from performance in 1966, except of course for the Bible, particularly on the 1967 album *John Wesley Harding*. Obviously, that became more specific and more evident in 1979, when Dylan's explicitly Christian phase began, but the Bible—Old Testament and New—is a text that was in Dylan's blood, both from his Hebrew school days in Hibbing and from growing up in that predominantly Christian community. The Christmas polkas and other Yuletide songs on Dylan's 2009 album, *Christmas in the Heart,* come from various musical traditions, but also from a place that is real, the mining town of Dylan's youth. Perhaps more important, since boyhood he had been listening to and absorbing gospel music and the blues on the radio, and this material, refracted through his own genius and his own writing, has always been part of his arsenal. In *Tangled Up in the Bible,* Michael J. Gilmore, who discusses intertextuality in his introduction, has collected many of the specific biblical intertexts.

Through the 1980s and early '90s there were plenty of great songs, and a couple of great albums, but most, Dylan included, would agree that with the release of his 1997 album, *Time Out of Mind,* came the dawning of the third classic period. The aesthetics of the songs are more intense, they are more up to date and timeless, no matter how old some of the material they are building on. With the folk and blues cover albums of 1992 (*Good As I Been To You*) and 1993 (*World Gone Wrong*), Dylan went back to school and returned to the blues and folk traditions from which he had so brilliantly diverted in the mid-1960s, when he made folk uncool, but which were always a part of him and whose strains would turn up in so many of the songs of *Time Out of Mind*.

This comeback for Bob Dylan has now lasted twenty years, in touring and recording, and in this period he again found his intertextual voice, embarking on a new mode of songwriting that has given his work a more conscious allusive and literary focus. This may have been what Dylan was hinting at in 2004, when Ed Bradley interviewed him for the CBS show *60 Minutes* following publication that year of *Chronicles: Volume One.* At one point Bradley asks Dylan if he could again write the songs like those from the 1960s, like "It's Alright, Ma (I'm Only Bleeding)." "Uh-uh," he replies, at which point Bradley asks: "Does that disappoint you?" Dylan follows up in cryptic fashion: "Well, you can't do something forever. I did it once, and I can do other things now. But, I can't do that."

The title *Time Out of Mind* is itself allusive or intertextual. It seems to borrow from, or at least allude to, singer-songwriter Warren Zevon's 1978 song "Accidentally Like a Martyr," the refrain of which ends with the phrase in question: "Never thought I'd ever be so lonely / After such a long, long time / **Time out of mind**." Zevon's title, "Accidentally Like a Martyr," alludes in turn to Dylan's song titles from 1965 and 1966 containing adverbs: "Positively 4th Street," "Queen Jane Approximately," "Absolutely Sweet Marie," "Obviously Five Believers." The Zevon song itself, "Accidentally Like a Martyr," along with the same artist's song "Mutineer," were regularly on Dylan's setlists during his fall tour of 2002, when Zevon was dying of cancer, as was public knowledge.

"Time Out of Mind" is also the title of and in the lyrics to a song by the jazz-rock group Steely Dan from their 1980 album, *Gaucho.* More important, the meaning of the phrase also points to time immemorial, time beyond memory, signaling a new phase for Dylan, in which he began to explore the past, eventually the very distant past, and conflate it with his own present, so producing worlds that are hard to pin down, and complex in the stories and images they conjure up. Dylan's album titles that followed show this process continuing. *"Love and Theft,"* the only Dylan album title in quotation marks, points to the intertextual thefts that album was carrying off. *Modern Times* puts in play the question of just how modern the times of the album will turn out to be. *Together Through Life* begs the question, "Whose life? Dylan's? The lives of fans who have been with him over the years?" And finally, *Tempest,* which ultimately comes from the Latin word for "time" (*tempus*), and whose worlds have an immense temporal range—Homer in the eighth century BC to John Lennon on Monday, December 8, 1980, when "they shot him in the back and down he went" ("Roll on, John"). These strands will all be brought together in the hybrid worlds built by Dylan's imagination, as he takes a line of Homer describing Odysseus "throwing filthy rags on his back like any slave" and gives it to Lennon, sharing with his old friend the identity with Odysseus that he took to himself, as we'll see on other songs on *Tempest*.

DYLAN UNLEASHED

Dylan provides the best discussion of what he is doing with these albums in the interviews he has given during this renaissance—cryptic as many of them are. Of central importance is Mikal Gilmore's interview "Dylan Unleashed," the one in which Dylan claimed, or seemed to claim, to have undergone "transfiguration." Gilmore knows his subject as well any of the interviewers

Dylan has singled out over the years. They started out on the back patio of a Santa Monica, California, restaurant: "At moments I pushed in on some questions, and Dylan pushed back. We continued the conversation over the next many days, on the phone and by way of some written responses." Dylan clearly wanted to get the story, and the message, straight. What followed was a commentary on his art, particularly in performance:

> Well . . . the *Time Out of Mind* record, that was the beginning of me making records for an audience that I was playing to every night. They were people from different walks of life, different environments and ages. There was no reason for these new people to hear songs I'd written 30 years earlier for different purposes.

And a little later again in the context of performance:

> Most of the songs work, whereas before, there might have been better records, but the songs don't work. So I'll stick with what I was doing after *Time Out of Mind,* rather than what I was doing in the seventies and eighties [he doesn't include the sixties], where the songs just don't work.

The setlists that followed in his performances through the rest of the fall of 2012 bear out what Dylan is saying here.

In these concerts there is a fairly even balance between, on the one hand, material from the 1960s up to *Blonde on Blonde* (1966), and on the other, songs from *Time Out of Mind* and later. He included "Tangled Up in Blue" regularly, as always, along with another occasional few from *Blood on the Tracks* (1975). By the following spring the songs of *Tempest* had taken the primary position as his setlists started to take on a more fixed quality. With a few notable exceptions, Dylan's songs from the late 1970s to the early 1990s are absent. They don't work in performance, and by now Dylan may have come to sense that with what had become of his art, the newer songs were simply much better.

At the end of the interview Gilmore takes up the issue of intertextuality in the songs:

> Before we end the conversation, I want to ask about the controversy over your quotations in your songs from the works of other writers. . . . Yet in folk and jazz, quotation is a rich and enriching tradition. What's your response to those kinds of charges?
>
> *Oh, yeah, in folk and jazz, quotation is a rich and enriching tradition. That certainly is true. It's true for everybody, but me. I mean, everyone else can do it but not me. There are different rules for me. . . . And if you think it's so easy to quote him and it can help your work, do it yourself and see how far you can get. Wussies and pussies complain about that stuff.*

It's an old thing—it's part of the tradition. It goes way back. These are the same people that tried to pin the name Judas on me. Judas, the most hated name in human history! If you think you've been called a bad name, try to work your way out from under that. Yeah, and for what? For playing an electric guitar? As if that is in some kind of way equatable to betraying our Lord and delivering him up to be crucified. All those evil motherfuckers can rot in hell.

Seriously?

I'm working within my art form. It's that simple. I work within the rules and limitations of it. There are authoritarian figures that can explain that kind of art form better to you than I can. It's called songwriting. It has to do with melody and rhythm, and then after that, anything goes. You make everything yours. We all do it.

In my view, the process of "transfiguration" that Dylan explores in this interview is more or less the process that literary critics call "intertextuality," perhaps an intertextuality of characters in the song, as much as the straight texts themselves. The process of transfiguration has been with Dylan from the beginning, from banging out Little Richard songs on his Hibbing High piano with a Little Richard hairdo, to channeling Woody Guthrie at the folk scenes of Dinkytown in Minneapolis, to going electric even before the appearance at Newport in 1965, and to all the phases that followed.

"TRYIN' TO GET TO HEAVEN"

Dylan's transfigurations in his dark and beautiful album *Time Out of Mind* find their roots in figures from the past, who take on new life thanks to the gift he was born with and the work to which he has put that gift. The singer in the first line on the first track of the album is "walking through streets that are dead, / Walking, walking with you in my head." Nothing is resolved by the time the song, "Love Sick," ends: "Just don't know what to do / I'd give anything to be with you." The images in the album are of humans isolated in a world of trouble. The melancholy is only made more stunningly poignant and beautiful by the bluesy voice of the singer and the music of it all, even if Dylan himself felt producer Daniel Lanois overdubbed and rearranged the album in a way that distanced it from the effect he himself was going for in the studio.

The narrator has other company in these songs, in the characters that are part of the intertextual fabric of some of the lyrics. "Tryin' to Get to Heaven" is such a song. When it came out, Dylan fans, myself included, participated in a collaborative effort to identify all the blues, folk, and gospel voices that haunt it and form its backdrop. "Tryin' to Get to Heaven" has at least ten intertexts that Dylan arranges and reworks to produce a song whose elements speak from their own original contexts, while at the same time becoming integral and vital parts of the new song. We might ask, what effect do these intertexts have in the song? How does their presence contribute to the mood and meaning of what is being sung? Here is the song, with the allusions, or intertexts, noted as they occur (Dylan's lyrics are in bold, with the corresponding intertext italicized below):

The air is gettin' hotter,
There's a rumblin' in the skies.
I've been wadin' through the high muddy waters,
I wade muddy waters, trying to reach dry land
—Tom Rush, "Turn Your Money Green"

But the heat riseth in my eyes.
Everyday your memory goes dimmer,
It doesn't haunt me like it did before.
I've been walkin' through the middle of nowhere,
Tryin' to get to heaven before they close the door.
Look at that sister comin' 'long slow,
She's tryin' to get to Heaven fo' they close the do'
—Alan Lomax, "The Old Ark's A-Moverin'," *The Folk Songs of North America*

When I was in Missouri,
They would not let me be.
I had to leave there in a hurry,
I only saw what they let me see.

I was in Missouri they would not let me be
Yeah when I was in Missouri, baby, would not let me be
No and I could not rest content until I come to Tennessee

—Furry Lewis, Tom Rush, et al.,
"Turn Your Money Green" (first and last verse)

You broke a heart that loved you,
You have wrecked a heart that loved you

—Byron Arnold, "Golden Chain," *Folk Songs of Alabama*

Now you can seal up the book and not write anymore.
Seal up your book, John,
An' don't write no more,
O John, John,
An' don't write no more.

—Alan Lomax, "John the Revelator," *Our Singing Country*

I've been walkin' that lonesome valley,
You got to walk that lonesome valley . . .
Jesus walked that lonesome valley

—"Lonesome Valley," African American spiritual

I've been walking that Lincoln highway

—Woody Guthrie, "Hard Travelling"

Tryin' to get to heaven before they close the door.
She's tryin' to get to Heaven fo' they close the do'

People on the platforms,
waitin' for the trains.
I can hear their hearts a-beatin',
like pendulum swingin' on chains.
One foot is on the platform
The other one on the train,
I'm going back to New Orleans
To wear that ball and chain

— Alan Lomax,"The Rising Sun Blues,"
Folk Songs of North America

I'm standing on a platform
Smoking a big cigar
Waiting for some old freight train
Carrying an empty car

—Woody Guthrie, "Poor Boy"

When you think that you've lost everything,
You find out you can always lose a little more.
I'm just going down the road feelin' bad,
I'm going down the road feeling bad
I'm going down the road feeling bad

I'm going down the road feeling bad, Lord, Lord
And I ain't gonna be treated this-a-way.
—Woody Guthrie, "Going Down the Road Feeling Bad"

Tryin' to get to heaven before they close the door.
She's tryin' to get to Heaven fo' they close the do'
I'm goin' down the river,
Down to New Orleans.
They tell me everything is gonna be all right,
But I don't know what "all right" even means.
I was ridin' in a buggy with Miss Mary Jane,
Ridin' in the buggy, Miss Mary Jane
Miss Mary Jane, Miss Mary Jane (twice)
—Alan Lomax, "Miss Mary Jane," *The Folk Songs of North America*

Miss Mary Jane got a house in Baltimore.
Sally got a house in Baltimo',
Baltimo', Baltimo'
Sally got a house in Baltimo'
An' it's full of chicken pie
—Alan Lomax, "Miss Mary Jane," *The Folk Songs of North America*

I've been all around the world boys,
I'm tryin' to get to heaven before they close the door.

She's tryin' to get to Heaven fo' they close the do'

Gonna sleep down in the parlor,
And relive my dreams.
I close my eyes and I wonder,
If everything is as hollow as it seems.
Some trains don't pull no gamblers,
No midnight ramblers like they did before.
This train don't carry no gamblers, this train,
This train don't carry no gamblers, this train,
This train don't carry no gamblers
No hypocrites, no midnight ramblers,
This train is bound for glory, this train.

—Woody Guthrie, "Bound for Glory"

I've been to Sugartown, I shook the sugar down,
I wanted su-gah very much,
I went to Sug-ah Town,
I climbed up in that sug-ah tree
An' I shook that sug-ah down.

—Byron Arnold, "Buck-Eye Rabbit,"
Folk Songs of Alabama

Now I'm tryin' to get to heaven before they close the door.
She's tryin' to get to Heaven fo' they close the do'

Dylan had done this sort of thing before, but never on such a grand scale. In "Tryin' to Get to Heaven" he steals from a wide range of songs, a good example of what he meant in the 2012 interview we saw with Gilmore, worth repeating: "It's called songwriting. . . . You make everything yours."

Nobody but Dylan could integrate such a disparate range of source texts, which he knows as part of the songbook that is in his head, into the compellingly poetic story that the song becomes. Dylan has famously said that in "A Hard Rain's A-Gonna Fall," every line could have been the first line of a song. Where the lines in "Hard Rain" had come from places that even he did not understand, the Muse, that is, his memory of the whole tradition, was now handing him scripts, which he integrated, orchestrated, and expanded into the song he made. Here in 1997, he has control of a dizzying variety of fragments of traditional gospel and folk songs and he knows what he wants to do with them. As the song is set out here you can read the intertexts, but what matters is what Dylan does with them. "I wade muddy water" becomes "I've been wadin' through the high muddy waters," thus providing a foretaste of the "High water risin'" that will open the apocalyptic song "High Water (For Charley Patton)" on Dylan's next album, *"Love and Theft."*

The resulting song is a powerful piece of writing, the story of someone in trouble, on the run it seems, who had to get out of Missouri, where "they would not let [him] be," now heading down to New Orleans. The biblical references to "that lonesome valley," and the refrain of the title, convey a sense that the end may be close—"Tryin' to get to heaven before they close the door." As the original details of the song accumulate, there is a heightening of the mysterious. Why wouldn't "they" let the narrator be back in Missouri, and why is he going down the river to New Orleans? Once we realize the components of the song and activate the contexts from which they come, the "I" of the song takes us with him, or her, or both, through the disparate worlds of those source texts, walking the lonesome valley like Jesus, going down the river bound for a version of the House of the Rising Sun, which has "been the ruin of many a young girl," in the words of that song, doing penance by sleeping down in the parlor, not upstairs with the clients.

Along with the echoes of those other songs, there is the poetry and songwriting of Bob Dylan, which ties it all together and further complicates the scenes. The singer is getting over a love lost and he is still hurting: "Every day your memory gets dimmer / It doesn't haunt me like it did before. . . . You broke a heart that loved you." As the song works toward its end there seems to be little cause for hope: "When you think that you've lost everything / You find out you can always lose a little more." The hope lies in the beauty of the song itself, the perfection of the writing and the tightness and progression of the verses, compensation here for the fact that the singer, as conveyed by the voice of Dylan, may not have too much time left before they close the door.

"HIGHLANDS"

Was Robert Burns, the great Scottish poet and songwriter, transfigured? Did he in fact become the singer of "Highlands," the dramatic and narrative reflection that closes *Time Out of Mind*? At 16 minutes and 31 seconds, it is Dylan's longest song—though in the few live performances of the song, Dylan sped it up and regularized the tempo, and cut some lines, getting it down to around 10 minutes. As he surely meant us to see, "Highlands" owes an obvious debt to Burns's "My Heart's in the Highlands." As with other intertexts, this is meant to be noticed right from the beginning of the song:

Well my heart's in the Highlands gentle and fair
Honeysuckle blooming in the wildwood air
Bluebells blazing where the Aberdeen waters flow
Well my heart's in the Highland
I'm gonna go there when I feel good enough to go

Burns's poem is only four verses long:

My heart's in the Highlands, my heart is not here;
My heart's in the Highlands a-chasing the deer;
A-chasing the wild-deer, and following the roe,
My heart's in the Highlands wherever I go.

It's a simple sentimental song-poem that is melancholic in its dwelling on memory and the absence of the object, but it hardly rises to a level of aesthetic beauty or meaning that has the power to affect us profoundly. Its AABB rhyme, the simplicity of the repeated frames, and the lack of any profound thought keep it on an unsophisticated level. Dylan found it, took what he wanted, and discarded the rest. But in the process, he tied himself to the folk tradition in which Burns was writing, a tradition within which Dylan himself has always been working, even before 1962, when at the Gaslight in New York City he sang a beautiful version of "Barbara Allen," the traditional Scottish ballad that Burns would have known almost two hundred years before as "Bonny Barbara Allen"—already an old song for Burns.

The Burns poem comes across strongly at one point in Dylan's song, fifteen verses in:

Well my heart's in the Highlands, with the horses and
hounds
Way up in the border country, far from the towns
With the twang of the arrow and a snap of the bow

Dylan's verbal debt to Burns is in fact fairly slight, just the five or six words with which he and Burns both begin ("my heart's in the Highlands"), and he has almost deliberately avoided further exact intertexts, replacing the targets of the Burns hunt (deer, roe) with its instruments (horse, hounds, arrows, bow). However, because of those opening words, and because of the geographical specificity ("Aberdeen waters"), the presence and mood of Burns's poem is strongly felt, then transformed by the poetry of Dylan, who talked a bit about how parts of the album came together in a 2001 interview with Mikal Gilmore for *Rolling Stone:* "I'd been writing down couplets and verses and things, and then putting them together at later times." Could Dylan have been aware of something very similar that Burns said about the process of writing his own song, "My Heart's in the Highlands," something I came across in a commentary on the songs of Burns? In Burns's manuscript, "Notes on Scottish Song," published in 1908, he wrote, "The first half-stanza [the five words in question] of this song is old; the rest is mine." These words could easily have come from a Dylan interview, more proof that what he is doing is old, conscious, and fully deliberate, and in a tradition, taking what you need from wherever you find it and using it to create new art.

The structure of Dylan's song is intricate and carefully devised, with the verses advancing the story of the song: chorus: 2 verses: chorus: 2 verses: chorus: 7 verses: chorus: 4 verses: chorus. In all, excluding the choruses, there are four verses on either side of the seven-verse-long scene, the heart of the poem. Those seven verses, beginning "I'm in Boston town, in some restaurant," form a dreamlike narrative centerpiece that is a song in itself.

In each of the choruses, as in the one cited, "My Heart's in the Highlands" is repeated from the first to the fourth line, while the second and third lines capture images from the beauty of the idyllic world held in the heart of the singer: "honeysuckle blooming," "bluebells blazing," "the wind, it whispers to the buckeyed trees in rhyme," "big white clouds like chariots that swing down low," "way up in the border country far from the town," "where the Aberdeen waters flow," and "by the beautiful lake of the Black Swan." Each chorus ends with a fifth line speculating on when or how his heart and he will be united in the Highlands, the first four as follows:

> I'm gonna go there when I feel good enough to go (1)
> I can only get there one step at a time (4)
> Only place left to go (7)
> Can't see any other way to go (15)

These idyllic scenes and the poetic colors with which Dylan paints them are all in stark contrast to what is going on in the life of the speaker. He seems in the first verse to be waking from a nightmare in a life of tedium and repetitive despair, looking "at the same old page / Same ol' rat race / Life in the same ol' cage." Part of the genius of Dylan's art in this new phase lies in his ability to establish voices, or sets of voices, of characters who use language that is old, archaic even. Sometimes that language is centuries old, sometimes decades old. This is a literary and musical development that has the brilliant effect of alienating the singer and setting him in a time that is unfamiliar and distant from that of any contemporary reality: "Strumming on my gay guitar," off in "gay Paree," "Dixie bound," and in "Boston town."

So in "Highlands" "same ol' rat race" has a 1960s feel to it, situating the speaker in a time that is no more, a world in which he doesn't fit. The third verse further brings out his alienation, a "prisoner in a world of mystery," using language, such as "Wouldn't know the difference between a real blonde and a fake," that suggests the mystery has to do with his incomprehension of what has been going on in the 1980s and '90s. "I wish someone would come / And push back the clock for me," he says. To when, 1965? Even further perhaps, to the time of Robert Burns? The narrator is "listening to Neil Young," as many of us still are, and he's "gotta turn up the sound," perhaps hard of hearing from too many loud concerts. Those listening to Dylan singing these lines may sympathize with his tastes, but the narrator's tastes are not shared by those in the world around him in the song: "Someone's always yelling turn it down." This is part of the isolation of the character that Dylan is constructing.

Six minutes into "Highlands," the song takes on the appearance of a dramatic interlude, a five-minute mini-play, verses 8–14 relating the experience in that Boston restaurant." Clinton Heylin finds fault with the scene:

> By then [at the recording session] Dylan had inserted a section of dialogue into the song—in which a series of non-sequiturs depict a run-in with a waitress in a New England bar-restaurant, stretching the song out while distancing it from its source.

Rather than being an insertion, I would maintain the dialogue is an integral part of the story Dylan is building in this song, a picture he is drawing of someone whom the world has passed by. The restaurant is empty, allowing the focus to be directed to the two players, the singer and the waitress: "Nobody in the place but me and her." Why is it empty? "It must be a holiday, there's nobody around," not particularly logical, and at the end of the exchange he steps "outside back to the busy street," so the emptiness inside is part of the design. There is a sense of attraction, or at least of appreciation—"She got a pretty face," perhaps, he fancies—reciprocated: "She studies me closely as I sit down."

This has happened before, in another lifetime, in reality perhaps thirty years earlier, before that waitress in Boston town was born even, a moment dramatized twenty-two years earlier in Dylan's most famous song from the 1970s, "Tangled Up in Blue," in a scene for which the singer of "Highlands," or Bob Dylan if we want, has been searching:

She was workin' in a topless place
And I stopped in for a beer
I just kept looking at the side of her face
In the spotlight so clear
And later on as the crowd thinned out
I's about to do the same
She was standing there in back of my chair
Said to me, "Don't I know your name?"
I muttered somethin' underneath my breath
She studied the lines on my face
I must admit I felt a little uneasy
When she bent down to tie the laces of my shoe
Tangled up in blue

The past encounter recalled in "Tangled Up in Blue" has become the present of "Highlands," or that's what he hopes. Maybe she'll pick him out again? The girl back then took him to her place, "lit a burner on the stove," and got high with him, taking the lead: "I thought you'd never say hello," she said. "You look like the silent type." She was first to speak back then, but now, at least on the original recording of "Highlands," he opens things up:

She got a pretty face and long white shiny legs
I say "Tell me what I want"
She says "You probably want hard-boiled eggs"

The exchange continues, still strangely,

I say "That's right, bring me some."
She says "We ain't got any, you picked the wrong time to come."

Then why did she bring them up? This exchange may seem odd, intensely surreal, even in the official version in *Bob Dylan: The Lyrics: 1961–2012:*

She got a pretty face and long white shiny legs
She says, "What'll it be?"
I say "I don't know, you got any soft-boiled eggs?"
She looks at me, says "I'd bring you some
But we're out of 'm, you picked the wrong time to come."

But he has a purpose here. "Soft-boiled eggs" is blues slang, a reference to part of the female anatomy, rightly rhymed here with "long white shiny legs." Dylan surely knows any number of versions of the blues song "Big Fat Woman," which even made it onto blues and folk artist Tom Rush's 1963 album, *Blues, Songs & Ballads:*

She's a fine lookin' woman, got great big legs
Big Fat Woman got great big legs
Ev'ry time she moves, move like a soft-boil'd egg

But in the Boston restaurant in the 1997 song, with that waitress on "Highlands," that sort of innuendo wasn't going to work. Unlike the woman who was "workin' in the topless place" in "Tangled Up in Blue," this one again reminds him of where he is in time: "you picked the wrong time to come," she says. This is 1997, not 1965.

Undeterred, he keeps trying. Recognizing that he is an artist, she asks him to draw a picture of her, to which he replies, "I would if I could, / but I don't do sketches from memory," a suggestion that he would do so back at his place, a version of the early- to mid-twentieth-century pickup line, largely a joke, "come up and see my etchings." She doesn't get it and wants him to do it on the spot: "I'm right in front of you, or haven't you looked?" As the exchange continues, he further pleads, "I don't know where my pencil is at," but the waitress produces one from behind her ear, and he proceeds to draw a few lines and "shows it for her to see." To her claim that it doesn't look a bit like her, he responds, "Oh kind miss, it most certainly does," adding to her reply, "You must be joking," his own rejoinder "I wish I was." Who did it look like? Perhaps again that other woman who was working in the topless place back in time, perhaps even Sara, onetime Playboy Club Bunny, in the song of the same name that he sang more than twenty years before, lingering from the last line of the album *Desire:* "Don't ever leave me, don't ever go."

Meanwhile, back in the restaurant in Boston town on "Highlands" and in 1997, things come to a head and the exchange is about to break down as the feminist waitress with the "pretty face and long white shiny legs" changes the topic in a way that will clinch the case that he is in the wrong time and place:

Then she says, "You don't read women authors, do you?"
At least that's what I think I hear her say
"Well," I say, "how would you know and what would it matter anyway?"
"Well," she says, "you just don't seem like you do!"
I said, "You're way wrong"
"Which ones have you read then?" I say "I read Erica Jong!"

And so the character of "Highlands" is stuck in the past with his Neil Young and his Erica Jong. For him Erica Jong is the author of the 1973 blockbuster *Fear of Flying*, in which Jong coined the phrase "zipless fuck" to capture her ideal of the perfectly liberated woman's right to have sex with strangers. But it didn't work on the waitress in the song, who without a word of response at this point leaves the stage.

Dylan generally doesn't play his longer songs in concert. "Desolation Row" (11:20) is an exception, though it is always abbreviated. He has never performed "Sad-Eyed Lady of the Lowlands" (1966) or "Tempest" (2012), long songs but shorter than "Highlands." As I've mentioned, singing at a faster tempo, Dylan was able to trim "Highlands" to around ten minutes to perform it, and I want to draw attention to the first two concerts at which he played it, a month apart in 1999, eighteen months after the release of *Time Out of Mind*. In both concerts (Chula Vista, California, on June 25 and Madison Square Garden in New York on July 27) it came in the setlist right after "Tangled Up in Blue," the song to which I am suggesting it served as a sequel, offering as it did an intricate intertextual response. Dylan performed the song precisely so he could have it back-to-back with "Tangled Up in Blue," a glimpse of the then-and-now across a quarter of a century.

After the waitress walks away in "Highlands," he steps back out into the street and, with a chorus closing the restaurant scene, gives us a commentary on what has just happened: "Some things in life, it gets too late to learn." The sense of alienation continues in the four verses that follow, but he recovers by the end of the song, realizing that the Highlands he's been looking for is not in any place or time, that is, not in any time out of his own mind:

Well, my heart's in the Highlands at the break of day
Over the hills and far away
There's a way to get there and I'll figure it out somehow
But I'm already there in my mind
And that's good enough for now.

And so the album ends, with the singer's expression of contentment with the way he is. In this song, the last one on an album that would prove to be the beginning of a new period of now classic songs, albums, and performance, stretching across the next twenty years, Dylan creates a voice whose songs look back across the years from a place of artistic confidence. These last words on the album *Time Out of Mind* reveal an artist who has rediscovered his genius. Leading into this verse, the singer makes a discovery: "I got new eyes / Everything looks far away." Far away in space and in time, back into the past but also into the future.

Or, as he would put it in "Bye and Bye," on the 2001 album *"Love and Theft,"* "The future for me is already a thing of the past." The quote marks around the title of this, the thirty-first studio album, are unique to the album, indicating that even the title is "stolen" and further pointing to the new process of theft, plagiarism, or intertextuality that is at the very center of his art, particularly the songs released in the new millennium.

"SHADOWS ARE FALLING": DYLAN'S MAJESTIC SADNESS

"Not Dark Yet," the seventh track on *Time Out of Mind,* from 1997, is among the most poetic of Dylan's songs. The weary but strong voice of the singer is closer to his end than his beginning, the opening three words, "Shadows are falling," finding their response in the close of each of the four verses: "It's not dark yet, but it's getting there." In the line "Behind every beautiful thing there's been some kind of pain," the apparent paradox captures the sound of the song, which in the moment of its being sung describes with beauty the very despair the song constructs. Each of the four verses fills in the details of the life of someone who has been through much, suffered much, getting ready for the dark, but still has hope, even if the last lines suggest it is not quite at hand: "Don't even hear the murmur of a prayer / It's not dark yet but it's getting there."

It is on *Time Out of Mind* that Dylan's aesthetics of melancholy are at their darkest, "bleak and riveting," as critic Jon Pareles put it, "closer than ever to the clear-eyed fatalism of classic blues." A number of the songs, particularly as they close, confront love that is lost but can't be forgotten, the pain left hanging there as the songs end, frequently with the words of the song title:

"Love Sick"
"Just don't know what to do / I'd give anything just to be with you"

"Standing in the Doorway"
"You left me standing in the doorway crying / Blues wrapped around my head"

"Million Miles"
"Yes, I'm tryin' to get closer but I'm still a million miles from you"

"'Til I Fell in Love with You"

"I just don't know what I'm gonna do / I was all right 'til I fell in love with you"

"Make You Feel My Love"

"Go to the end of the world for you / To make you feel my love"

Melancholy as an aesthetic experience in any art form—whether music, literature, or painting—produces both pain and pleasure, the latter making the former bearable, compensating for it. It comes through contemplation of place and time, and frequently of persons lost or absent. It can be intensified by the memory of past events or situations now gone, and by the sense of loss that such memories evoke.

"Thou majestic in thy sadness at the doubtful doom of humankind." So wrote Alfred, Lord Tennyson in "To Virgil," a poem across the ages to the Roman poet who gave him so much. Tennyson would have had in mind a passage from the end of Virgil's *Georgics,* on Orpheus, the mythic poet numbered among those who "enriched our lives with the newfound arts they forged"—now including Dylan by way of the Nobel medal he has with that line of Virgil's Latin on it. Toward the end of the *Georgics,* Orpheus had through the power of his song rescued his wife, Eurydice, from the Underworld. But he made a human error and looked back, and she was lost, back into the pit:

Like smoke
Disintegrating into air she was
Dispersed away and vanished from his eyes
And never saw him again, and he was left
Clutching at shadows, with so much more to say
—Virgil, *Georgics* 4, tr. David Ferry

The majestic sadness of those lines comes from the same place that gives us the songs on *Time Out of Mind* and many since, including those from his recording of the Great American Songbook. The lyrical darkness of Dylan's vision of the last twenty years has brought him into a close alignment with Virgil. At the end of the song "Duquesne Whistle" (2012), written with Robert Hunter, we hear the singer wondering.

The lights of my native land are glowin'
I wonder if they'll know me next time around
I wonder if that old oak tree's still standing
That old oak tree, the one we used to climb.

Going into exile two thousand years earlier, Virgil's shepherd-poet wondered the same thing.

Oh, will it ever come to pass that I'll
Come back, after many years, to look upon
The turf roof of what had been my cottage

And the little field of grain that once was mine,
My own little kingdom.

—Virgil, *Eclogue* 1, tr. David Ferry

This ability to capture with compelling artistry the universal human emotion that comes with absence, and with memory of a place lost is just one reason Virgil and Bob Dylan both matter. It is time to put them more directly together.

7

MATURE POETS STEAL: VIRGIL, DYLAN, AND THE MAKING OF A CLASSIC

> Immature poets imitate; mature poets steal; bad poets deface what they take, and good poets make it into something better, or at least something different. The good poet welds his theft into a whole of feeling which is unique, utterly different from that from which it is torn; the bad poet throws it into something which has no cohesion. A good poet will usually borrow from authors remote in time, or alien in language, or diverse in interest.
>
> —T. S. Eliot, "Philip Massinger," 1920

A few days after it came out, on September 11, 2001, as mentioned earlier, I started listening to the songs on *"Love and Theft,"* and when I arrived at "Lonesome Day Blues," I heard Virgil, greatest of the Roman poets, singing with the voice of Dylan:

Dylan:

> *I'm gonna* ***spare*** *the* ***defeated****—I'm gonna speak to the crowd*
> *I'm gonna* ***spare*** *the* ***defeated****, boys I'm gonna speak to the crowd*
> *I am goin' to* ***teach peace*** *to the* ***conquered***
> *I'm gonna* ***tame the proud***

Virgil:

> remember Roman, these will be your arts:
> to **teach** the ways of **peace** to those you **conquer**,
> to **spare defeated peoples, tame the proud**

Spare the defeated, teach, peace, conquered, and tame the proud. This is beyond coincidence. Virgil's lines, from Book 6 of his epic, the *Aeneid,* are set in the Underworld. The ghost of Aeneas's father is instructing him, and future Romans, on how to conduct themselves as they build their empire, whose remains are still so visible today in the city of Rome. Aeneas will in fact fail at the end of the poem to live up to his father's instructions, as he kills his defeated enemy. That final move is a culmination of the second half of Virgil's poem, and of the epic wars of Aeneas, whose depiction is generally taken to allude to the civil wars of Julius Caesar and his adopted son, future emperor Augustus Caesar, whose propaganda presented the Caesars as the descendants of Aeneas, and so of Venus, the divine mother of Aeneas. It is this darker aspect of Virgil's poem that seems to be appealing to Dylan.

Vietnam, the war of Dylan's youth, also seems very much in the air in the words of the "Lonesome Day Blues," whose singer is in a bad way, stripped through death, desertion, and elopement of all his family: "My pa he died and left me, my brother got killed in the war / My sister she ran off and got married, never was heard of any more"; also in the sixth verse: "Set my dial on the radio / I wish my mother was still alive." In the official *Bob Dylan: The Lyrics: 1961–2012,* the second line reads, "I'm telling myself I'm still alive." So Vietnam is the natural setting for the song, at least as heard by any baby boomer with ears to hear. But once we recognize the Virgilian intertext and its context of the ancient Roman civil wars, something happens to the song's meaning. The two contexts, familiar to me—Rome and America—merge and make the song about no war and every war, as happens so often with Dylan's lack of specificity around time and place in his songwriting.

Of course this doubling of the temporal contexts proved to

be too simple, for there were other intertexts in the song, as the world of Dylanology discovered in the pages of the *Wall Street Journal* on July 8, 2003, almost two years after the album came out. An American named Chris Johnson had been browsing in a bookstore in Fukuoka, Japan, fifty miles from the town of Kitakyushu, where he was teaching English. On the first page of the English translation of a Japanese gangster novel, Junichi Saga's *Confessions of a Yakuza,* Johnson read "My old man would sit there like a feudal lord" and immediately recognized a line from Dylan's song "Floater (Too Much to Ask)," from the same album *"Love and Theft":* "My old man he's like some feudal lord." Over the next few days, the Dylan community responded with blog posts and articles on the popular site expectingrain.com, seeming to compete for the wittiest adaptation of Dylan's act of theft: "They ain't his, babe." "The lines they were a-changin'." "The Freestealin' Bob Dylan." But others also chimed in, defending Dylan's borrowings as cultural collage and traditional literary allusion, the way song and poetry have worked for centuries.

I was more interested in how the lifted passages might work in their new setting. A Japanese gangster novel and the Roman poet Virgil's *Aeneid,* side by side, felt like the sort of creative surrealistic juxtapositions that had its roots in Dylan's songwriting going back to the sixties, now with the disparate elements coming not from Rimbaud or folk song, though that would never disappear, but from literary texts, and specifically a Japanese novel. Dylan disperses some twelve undeniable passages from the novel across five songs on *"Love and Theft,"* generally two per song. This is a pattern that was to be repeated with the words of Ovid and Henry Timrod on *Modern Times* (2006), and Homer on *Tempest* (2012)—the presence of one allusion confirming the other on the song in question, showing the allusion to be no accident. This included two passages in "Lonesome Day Blues":

Dylan:

> *Samantha Brown* ***lived in my house*** *for about four or five months*
> ***Don't know how it looked to other people,***
> ***I never slept with her even once.***

Saga 208:

> Just because she **was in the same house** didn't mean we were living together as man and wife, so it wasn't any business of mine what she did. I **don't know how it looked to other people, but I never even slept with her—not once.**

Dylan:

> *Well my captain he's decorated—he's well schooled and he's skilled*

> *My captain, **he's decorated**—he's well schooled and he's*
> *skilled*
> *He's **not sentimental—don't bother him at all***
> ***How many of his pals have been killed.***

Saga 243:

> There was nothing **sentimental** about him—**it didn't bother him at all that some of his pals had been killed.** He said he'd been given any number of **decorations**, and I expect it was true.

Again, the context of the words is what matters, since Dylan, by quoting the words, is at least potentially invoking the situation those words describe. *Confessions of a Yakuza,* which Dylan probably picked up while on tour in Japan, is a remarkable piece of writing. It blurs the genres of novel and biography, fiction and nonfiction, a little like Dylan's own *Chronicles: Volume One*. Its narrative complexity, along with a lively use of colloquial language, must have appealed to Dylan's literary sensibilities. The author of *Confessions* is the doctor and novelist Junichi Saga, whose book tells the story of the life of an early- to mid-twentieth-century gangster, Ijichi Eijii (b. 1904), who was Saga's patient. The two stolen passages that appear in "Lonesome Day Blues" come from late in the novel. The first concerns a woman named Osei (= Samantha Brown), who is staying with Eiji not long before the American defeat of Japan in World War II.

The second quote from "Lonesome Day Blues" comes from Eiji's final narrative chapter, as he recalls running into Osei again in 1951: "The Korean War was going strong, and my new gambling place in Tokyo was doing really well." Why does this sound so much like a line from a Dylan song? The inspiration for Dylan's "decorated" captain is one Nagano Seiji, encountered while Eiji was in prison, and who had sliced off a fellow prisoner's arm. The pals whose deaths in Dylan's song "don't bother him at all" were the soon-to-be archenemies of the clearly American singer of "Lonesome Day Blues," Japanese soldiers who died in the Chinese-Japanese War (1937–45). History, reality, and fantasy are all put in the mix, with surrealistic effects that are only heightened as we see the texts Dylan is working into his song. "Tryin' to Get to Heaven" had drawn together blues, gospel, and folk to create its world. In "Lonesome Day Blues" Dylan repeats the method, but now with a Roman epic poem and Japanese novel providing the parts.

The war with which "Lonesome Day Blues" begins ("Well, my pa he died and left me, my brother got killed in the war") was about to become even more complex. Sometime in the early 2000s the critic and musician Eyolf Østrem pointed out on his blog that Dylan also layers two passages from Mark Twain's *The Adventures of Huckleberry Finn* in "Lonesome Day Blues":

Dylan:

> *My sister, she ran off and got married / Never was heard of any more.*

The Adventures of Huckleberry Finn, Ch. 16:

> . . . and my sister Mary Ann run off and got married and never was heard of no more . . .

Dylan:

> *Last night the wind was whisperin' somethin' —I / was trying to make out what it was / I tell myself something's comin' / But it never does*

The Adventures of Huckleberry Finn, Ch. 1:

> I felt so lonesome I most wished I was dead. The stars was shining, and the leaves rustled in the woods ever so mournful . . . and the wind was trying to whisper something to me and I couldn't make out what it was.

So now Dylan's song adds to the mix the writing of Mark Twain, with whom Dylan has always shared an affinity, up and down the Mississippi, on which they were both, near enough, born and raised. The first of the Huck Finn quotes comes from the Grangerford-Shepherdson episode of the novel. Huck quotes Buck Grangerford's explanation of the origins of the feud that is systematically eliminating the two families, all with the look of the Civil War, which Twain had lived through as a young man in his twenties:

> "Well," says Buck, "a feud is this way. A man has a quarrel with another man, and kills him; then that other man's brother kills *him;* then the other brothers, on both sides, goes for one another; then the *cousins* chip in—and by and by everybody's killed off, and there ain't no more feud." (Ch. 18)

Dylan's Twain quotes in "Lonesome Day Blues" complicate our understanding of the line "my brother got killed in the war," making us gather up all the references, and allusions, from Vietnam back to the American Civil War, with the borrowed Virgil lines taking us even further back in time, to the civil wars that tore apart the Roman republic, long an interest of Bob Dylan.

Dylan would give us some insight into what he was trying to do in "Lonesome Day Blues" and in other songs on the album in a *Rolling Stone* interview with Mikal Gilmore three months after its release:

> The whole album deals with power . . . the album deals with power, wealth, knowledge and salvation. . . . It speaks in a noble language. It speaks of the issues or the ideals of an age in some nation, and hopefully, it would also speak across the ages.

The generalized language of Dylan's description "a noble language," "of an age in some nation," "across the ages," fits well with the voices and texts that are activated in "Lonesome Day Blues." These nations and ages include imperial Rome and imperial Japan, but always America, particularly America of the nineteenth century, a time in which many of Dylan's lyrics, even he himself, have taken up residence. These include the world of Rome, the world in which Virgil saw Augustus, divine descendant of the hero Aeneas, turn republic into empire. Nothing could better suit historical reality than what Dylan sings in "Bye and Bye":

> *I'm gonna establish my rule through civil war*
> *Gonna make you see just how loyal and true a man can be.*

Again, in another of the songs on the album, "Honest with Me":

> *I'm here to create the new imperial empire*
> *I'm going to do whatever circumstances require*

The dark places from which these lines come stretch back "across the ages," to a poet whose status within his own culture has much in common with that of Dylan in our times. Creative genius can emerge in human history at any time and in any place. When that happens, similarities may emerge, accidentally or by design.

A TALE OF TWO "PLAGIARISTS": VIRGIL AND DYLAN

In Chapter 5 we saw Virgil's intentional borrowing from Homer, with Odysseus's encounter with his mother woven into the fabric of the *Aeneid* and its hero's meeting with the shade of his wife and father. There was no attempt to conceal the theft; yet again, a reader's noticing the theft and activating the meaning of the stolen lines is part of what makes new meaning. Two thousand years ago critics noticed, as the Roman historian Suetonius (c. 69–122), recorded in his *Life of Virgil:*

> Asconius Pedianus in a book he wrote "Against the Detractors of Virgil," sets out a few of the charges against him, dealing with historical detail, and with the accusation that he took many lines from Homer. He reports that Virgil would defend himself against the accusation: "Why don't they try the same thefts? If they do they'll find out it's easier to steal Hercules' club from him than to steal a line from Homer."

Some of them got it, some didn't. This is what T. S. Eliot meant when he wrote, "Immature poets imitate; mature poets steal." Mature poets make the old line new, and make it their own, improve on it, or at least match it. Dylan, who would come to steal from Homer in the song "Early Roman Kings," seemed to be channeling Virgil's response to his critic in the interview with Mikal Gilmore in 2012 following the release of *Tempest:*

> And if you think it is so easy to quote him [Henry Timrod] and it can help your song, **do it yourself and see how far you can get**. It's an old thing—it's part of the tradition. It goes way back.

Dylan is no longer just talking about the folk tradition, about "Barbara Allen" going back to the time of the Renaissance, or songs on *Modern Times* going back to Henry Timrod and the nineteenth century. Dylan here even seems to be stealing Virgil's response to his critics; more likely both are giving independent glimpses into how the intertextual process works, in any place or time. In the parallel voices of Virgil ("let them try it") and Dylan ("see how far they can get"), identical in tone, you hear the confidence of two mature poets, late in their careers, who know what they are doing, have always known what they are doing, and who are confident in the classical status that their genius and their art have achieved.

WHEN A POPULAR SONG BECOMES A CLASSIC

Virgil's works, like Dylan's song or Seamus Heaney's poetry, were taught in his own lifetime. That is one sign of genius, a recognition that something unusual is going on, and that the art of today is going to be around for many tomorrows, and is therefore worth introducing into the curriculum. That is historically unusual; it generally takes the passage of a lifetime or more for the curriculum to acknowledge that new art is worth teaching. Like Dylan, Virgil too came from a backwater region in the north, but eventually, also because of his song and his brilliance, found his way to the metropolis. Rome was not so much to his liking but that was where the action was, so that was where he went, at least for a time. He preferred the climate and the culture of Naples or Sicily, as Dylan would prefer that of upstate New York and then California, both away from Rome and New York City, the two capitals of the world two thousand years apart, in which they had earned their fame. Virgil also came from humble origins, perhaps the son of a potter, maybe a beekeeper. He died a wealthy man, worth over the equivalent of $10 million, presumably reward for his poetry, if we believe the ancient biography of Suetonius, written more than a century after the poet's death. In short, Virgil was a rock star in his time, as popular as any of the hugely popular lyre-singers from back in Homer's day and beyond. The historian Tacitus (c. 56–120), another favorite of Dylan's, describes Virgil's popularity as being

> vouched for by the letters of Augustus, and by the behavior of the citizens themselves; for on hearing a quotation from Virgil in the course of a theatrical performance, they rose to their feet as a man, and did homage to the poet, who happened to be present at the play, just as they would have done to Augustus himself.

Anyone who has attended a Dylan concert, in their seat until the opening cadences of "Tangled Up in Blue," can relate. Virgil was also said to be shy, or perhaps fearful of his fortune and fame, fearful even of his fans, as Suetonius reports:

> whenever he appeared in public in Rome, where he very rarely went, he would take refuge in the nearest house, to avoid those who followed and pointed him out.

Like Dylan, he was also widely "covered," by singers whose names do not survive: the success of the *Eclogues* on their first appearance was such that they were frequently performed onstage. These may even have been the first versions many Romans heard, as was true for many songs of Dylan, heard first in the performance of Joan Baez, Peter, Paul and Mary, the Byrds, and some even prefer those versions. The YouTube video of British pop singer Adele covering Dylan's "Make You Feel My

Love" on the David Letterman show in 2011 has had around 50 million hits. Only a million of us have seen the bootleg Dylan version of the same song from that show in Rome on November 6, 2013. I know which one I prefer.

We think of Virgil, or at least those of us who think of him these days, as a classic. But before he became a classic, he was part of popular culture, just as opera was pop before time turned it, or some of it, into something different. Every classic starts out as popular; it is read, viewed, and heard by the people, because its music, words, or images touch something in us, express universals that are profoundly meaningful. Art of any sort will only be read or listened to by future generations if the generation in which it is produced recognizes it first—there are exceptions, the painting of Van Gogh, for example, who was ahead of his time, but people eventually got it in his case. When popular works continue to be meaningful beyond their time, they attain a status that can best be termed "classical."

Virgil was read and followed by later poets and educated Romans, who were acquainted with his verses from their early school days. But he was also read by the people in the city of Pompeii; in more than fifty places, his lines were scrawled on walls, a century after his death, preserved by the volcanic ash that buried the city. In AD 73 or 74 a Roman soldier taking part in the siege of the Jewish fortress at Masada wrote a line, preserved on a papyrus scrap, that Virgil's Carthaginian queen Dido spoke to her sister, "Anna, my sister, what dreams terrify me in my anxiety!" And at the other end of the Roman empire his lines are the only poetry to be found on wooden tablets excavated at Vindolanda, a fort on Hadrian's Wall in the north of England.

It is another feature of the classic that it creates tags, brief quotes that lend an air of authority. The user of the tag may at some point become unaware of the source, or at least the specific context of the quote, but tags both have a poetic quality and a message that is timeless, and so can have a life of their own, no longer tied to context. Shakespeare provides numerous examples, "to be or not to be," "pound of flesh," "a rose by any other name," and so on. Or T. S. Eliot's "Let us go then, you and I," "Do I dare to eat a peach," or "April is the cruelest month," and so on. Virgil generated his own share, most famously "beware of Greeks bearing gifts," and "Love conquers all." Dylan has joined this company. I recently emailed a colleague, no fan of Dylan to my knowledge, to tell him I wouldn't be able to make it to a conference, given the deadline for this book. His response, "Don't think twice, it's all right," showed that the Dylan tag had entered his consciousness, and emerged as a version of "that's okay." "The times they are a-changin'" and "the answer is blowin' in the wind," "only a pawn in their game," have similar status.

"WHAT IS A CLASSIC?"

It is no coincidence that Sir Christopher Ricks, the distinguished and prolific literary critic and editor of English literature, has produced the fullest editions, with variant readings duly recorded, of both T. S. Eliot (2015) and Bob Dylan (2014). In each case, it is possible to see from these editions slight and less slight changes in poems or songs that came about by changes of mind by the artist or by performance variation. The classic text is something that sticks around, that outlasts its moment. That is why, separate from Ricks's 961-page songbook, there have been three official editions of Bob Dylan lyrics, most recently in 2016: *Bob Dylan: The Lyrics: 1961–2012,* partly edited by the songwriter himself. If Dylan follows up *Tempest* with a new original album, posterity will need a fourth edition. Ricks works largely on the greatest English literature of the last five centuries, the seventeenth to the twenty-first. I work mostly on the five centuries from the third BC to the second AD, which gave us the best of Roman literature, works without which Dante, Milton, and Eliot would have written something, but something other than what they gave us. That Ricks and I come together on the art of Bob Dylan is no coincidence.

"What Is a Classic?" is an essay by T. S. Eliot, given as a lecture in London to the Virgil Society in 1944. Four years later, Eliot would himself be awarded the Nobel Prize in Literature. Eliot wrote this essay at a time of considerable crisis, a time when the end of the war in Europe seemed assured, but also a time at which it was not clear what the new Europe would look like, not unlike the present:

> We need to remind ourselves that, as Europe is a whole (and still, in its progressive mutilation and disfigurement, the organism out of which any world harmony must develop), so European literature is a whole, the several members of which cannot flourish, if the same blood-stream does not circulate throughout the whole body. The blood-stream of European literature is Latin and Greek—not as two systems of circulation, but one, for it is through Rome that our parentage in Greece must be traced.

Metaphors of European bloodstreams may seem odd, even vaguely disturbing, in the light of the anti-Semitism of some of Eliot's early poetry, but in thinking about Eliot it is important not to jettison the valuable with the odd. Eliot is also a poet who worked as Dylan would come to work, in his visions and revisions, in his being aware of writing in certain traditions. His thoughts on becoming a classic are therefore of some value. Almost everything in Dylan's early song can in some way be traced to the tradition of folk song and the blues. These are old traditions, and though they have been considered to be lower in register because of the social context of their performance, the greater simplicity of their melodies, or for any number of other reasons, with time they acquire a status that gives them a permanence, or "maturity" as Eliot put it.

A LIFETIME OF LABOR

I WROTE "BLOWIN' IN THE WIND" IN 10 MINUTES, JUST PUT THE WORDS TO AN OLD SPIRITUAL.

—BOB DYLAN, 2004

THAT IS ALL WE DID IN THOSE DAYS. WRITING IN THE BACK SEAT OF CARS AND WRITING SONGS ON STREET CORNERS OR ON PORCH SWINGS, SEEKING OUT THE EXPLOSIVE AREAS OF LIFE.

—BOB DYLAN, 1977

Early readers of Virgil were just as interested in how their poet put together his work as we are with Dylan or Eliot. So it is that details about the Roman poet's methods were preserved across 150 years and made it into the same *Life of Virgil:*

> When he was writing the *Georgics* it is said to have been his custom to dictate each day a large number of verses which he had composed in the morning, and then to spend the rest of the day in reducing them to a very small number, wittily remarking that he fashioned his poem after the manner of a she-bear, and gradually licked it into shape. In the case of the *Aeneid,* after writing a first draft in prose and dividing it into twelve books, he proceeded to turn into verse one part after another, taking them up just as he fancied, in no particular order. And so as not to check the flow of his

> thought, he passed over some things without finishing them and propped up others with very slight words, which he used to joke were put in like struts to support the work until the solid pillars arrive.

Dylan, in *Chronicles: Volume One,* which mentions the same Suetonius history of *The Twelve Caesars*, uses a similar metaphor for Hank Williams's songwriting (96):

> In time, I became aware that in Hank's recorded songs were the archetype rules of poetic songwriting. The architectural forms are like marble pillars and they had to be there.

Was Dylan talking about Hank Williams or about himself—or both? And was he also speaking across the centuries to Virgil, each working with the metaphor of pillars? Over the years we've been given occasional glimpses of how Dylan's writing and rewriting happens. Some of Dylan's songs may actually have come to him as easily as he would have us believe in interviews, but for the most part, his process has always been as painstaking and thoughtfully creative as that of any poet. There is now striking evidence from the new Bob Dylan Archive indicating common methodologies, shared across the centuries. Genius generally does not manifest itself spontaneously; it also takes hard work. The often-quoted statement from Romantic poet William Wordsworth, to the effect that "poetry is the spontaneous overflow of powerful feelings," should not be quoted in isolation from what follows: "it takes its origin from emotion recollected in tranquility." Great songs and poems may give the impression of being spontaneously and easily produced—that is the mark of a poem or song, that it comes across as perfect and as expressed the only way it could possibly have been expressed. It may be that some stage of the writing of "Blowin' in the Wind" did take just ten minutes, as he claims. But I have my doubts. As he said, "if you told the un-truth, well, that's still well and good."

In the *New Yorker* for October 24, 1964, Nat Hentoff recounts how Dylan recorded the entire album *Another Side of Bob Dylan* in a single session. "I've no idea what he's going to record tonight," recording producer Tom Wilson told Hentoff. "It's all to be stuff he's done in the last couple of months." As it happened, the recording was done in one session, fourteen songs from 7:05 P.M. till 1:30 A.M. the following day. Eleven would appear on the album, recorded at the rate of around two songs per hour. Hentoff talks about Dylan's lifestyle at this point:

> He prefers to keep most of his time for himself. He works only occasionally and during the rest of the year he travels or briefly stays in a house owned by his manager, Albert Grossman in Bearsville, New York—a small town adjacent to Woodstock and about

a hundred miles north of New York City. There Dylan writes songs, works on poetry, plays, and novels, rides his motorcycle, and talks with his friends. From time to time, Dylan comes to New York to record for Columbia Records.

There is the tranquility Wordsworth mentions as a necessary ingredient to the production of great poetry. And we can assume that for Dylan, as for any poet, there are interruptions and resumptions in the writing of the songs. One of the great ones on this album, "Chimes of Freedom," is a case in point. In February 1964, Dylan and three others went on a road trip from New York to New Orleans, then on to Denver and Berkeley. One of those on board, Dylan's tour manager, Victor Maymudes, claims that right from the start of the trip Dylan started writing "Chimes of Freedom," sitting in the backseat with a portable typewriter on his knee. We know he performed the song in Denver on February 14, a full eight months before the New York recording session. In 2005 a teaser emerged from Dylan's archive, a page of stationery from the Waldorf-Astoria hotel in Toronto, where Dylan was working on a TV program in late January 1964, before the road trip. The page is covered on both sides by handwritten lyrics of "Chimes of Freedom," with pen-and-pencil corrections. The song, which Dylan need not have written *during* the hotel stay itself, has most of its components, with each of the six stanzas set, though each still under construction with much not yet near its final form. The page refutes the claim that Dylan dashed off the poem or started composing it on a typewriter while riding in the back of a station wagon a month after the stay in Toronto. Whatever the truth of the matter, the writing of "Chimes of Freedom" took time and hard work to finalize, in addition to the innate genius of Bob Dylan.

The process continues in recording sessions, mostly with musical arrangement, but sometimes with lyrics, where Dylan has been known to rewrite during the sessions. Accompanying musicians wait while he jettisons poetry and rewrites lyrics. Now with the release in 2015 of take 1 of "Desolation Row" from *The Bootleg Series Vol. 12: The Cutting Edge 1965–1966,* we can get a sense of how this must go. Here the lyrics were more or less set before Dylan entered the studio, the product of his private toiling. But one exception comes in verse 7, which turns to Casanova, one of the presumed acquaintances whose faces the singer rearranges and disguises with such literary and fantastic names. The final version is just right:

They're spoonfeeding Casanova
To get him to feel more assured
Then they'll kill him with self-confidence
After poisoning him with words

But on take 1 things were very different. "Spoonfeeding" had a direct object, and its subject, rather than the ominous "they," the ones pulling the strings, is "he," the Phantom of the Opera: "He is spoonfeeding Casanova / Boiled guts of birds / He will kill him with self-confidence." In take 5 "He is spoonfeeding Casanova / The guts of birds / Then he'll club him with self-confidence." Eventually, after how many more takes we don't know, it gets to the right place. The "boiled guts of birds" are gone and the "spoonfeeding Casanova," without that direct object, has an ominous power of its own, its subject no longer the Phantom, rather the mysterious "they." Without the evidence of the earlier takes we would never imagine that it could ever have been otherwise. This is simply one small example and provides a glimpse into the creative rethinking, taste, and judgment that go into poetic perfection, the only thing that matters for the artist, be he Dylan or Virgil.

THE TULSA ARCHIVE AND "DUSTY SWEATBOX BLUES"

On March 2, 2016, the *New York Times* published an article by Ben Sisario, "Bob Dylan's Secret Archive," which reported that the George Kaiser Family Foundation had bought six thousand pieces, the entire archive of Bob Dylan—concert recordings, hours and hours of film, the leather jacket he wore at Newport in 1965, the night he went electric. The archive includes notebooks and scraps of paper, hotel stationery with cigarette burns and coffee stains, and of most interest for my purposes, Dylan's song drafts. It is now held in the Helmerich Center for American Research at the University of Tulsa, where it will remain available to Dylan scholars. The Bob Dylan Center is scheduled to open in 2019 in downtown Tulsa, and will display some materials rotating in and out from the archive. Kaiser went to Harvard and is three years younger than Dylan and Joan Baez. His generosity was partially motivated by what was happening in his college days in Cambridge: "I was taken by Joan Baez in college when she was singing down the block." The world of Dylan research has been transformed, and the results will show for years to come. Thanks to the generosity of Dylan's manager of nearly thirty years, music producer Jeff Rosen; Larry Jenkins, who has been coordinating with the archive on Dylan's behalf; and Michael Chaiken, inaugural curator of the archive, I can here give a glimpse, from images and transcriptions of Dylan's writing, into the archive and the working mind of Bob Dylan. Fasten your seat belts.

The Tulsa archive for the first time reveals a process something like that recorded for Virgil—in his case writing or reciting a number of lines in the morning, then going back, deleting, changing, licking into shape. In the archive there is one five-by-three-inch blue spiral notebook of forty-five pages, which once cost nineteen cents, as its cover announces. On its pages Bob Dylan worked in miniature handwriting on drafts of several of the songs of *Blood on the Tracks,* the classic album of 1975. None of the songs is complete, with some unrecognizably distant from their final perfection, proof that just like Virgil, "he proceeded to turn into verse one part after another, taking them up just as he fancied, in no particular order." Indeed, there are even lines that would end up in other songs: one line in a draft of "Tangled Up in Blue," "As I watched you disappear over that lonesome hill," would end up in the superb outtake "Up to Me," released on *Biograph* in 1985, more than ten years after it was written: "Well I watched you slowly disappear down into the officer's club." Traces of another draft line from "Tangled Up in Blue," "When you needed me most I was always off by myself," even ended up, changed, maybe less honest than the earlier one, on another album, *Desire* (1976), though addressed to the same woman, in the song "Sara": "You always responded when I needed your help."

Only some song titles are set. On one page "Idiot Wind" has the title "Selfish Child." And the iconic song of the seventies, sole survivor of that decade in the setlist of Dylan's 2017 tour, was once to be called "Dusty Sweatbox Blues," one of the most striking revelations from the Tulsa archive. Music critic and author Neil McCormick describes this song in his own virtuoso sentence:

> The most dazzling lyric ever written, an abstract narrative of relationships told in an amorphous blend of first and third person, rolling past, present and future together, spilling out in tripping cadences and audacious internal rhymes, ripe with sharply turned images and observations and filled with a painfully desperate longing.

It took genius, imagination, and hard work to get the song to that place. On page 23 of the notebook, we see the close of the first verse from back then. Fortunately, Dylan kept working at it and we did not get to hear this version:

And I was walking by the side of the road
Rain falling on my shoes
Headin' out to the old east coast
Lord knows I paid some dues
Wish I could lose, these dusty sweatbox blues.

"Tangled Up in Blue," as the song was renamed, that title also in the notebook, has featured in some 1,600 concerts, while the concert video of the song from the Rolling Thunder Revue tour of 1975–76 has had around 20 million views on YouTube. By now it may be the most recognized song in Dylan's huge arsenal. It is also a song that went through many changes, from the rejected 1974 New York session (which came out in 1991 on *The Bootleg Series Vol 1–3: Rare & Unreleased 1961–1991*) to the familiar Minnesota version on *Blood on the Tracks*. And it has had as many variations in performance as any Dylan song.

The title was eventually changed, possibly "Blue Carnation," then "Tangled Up in Blue" as revealed on page 15. At the head of the page before that there is a list of possible words for ending the second-to-last line of each verse, which all end with the song's title, "tangled up in blue": "Jew–who–few clue–

do–flew–grew–new–rue–sue–too–you–zoo–shoe–glue–view." Five of these turn up on the song, along with two new ones, "through" and "avenue."

The woman of "Tangled Up in Blue," already in the singer's past when he is "layin' in bed / Wond'rin' if she'd changed at all / If her hair was still red," is naturally taken to have some basis in Dylan's wife Sara Dylan, who had worked as a "Bunny" in the Playboy Club in New York before Dylan met her: she "was working in a topless place / when I stopped in for a beer," as the song puts it. Across a number of pages of the notebook her role shifts and refracts as Dylan seeks to find just how he wants to describe the thinly camouflaged Sara, "so easy to look at, so hard to define," in the song "Sara," from *Desire*. In the notebook, with a tiny difference—"she was *dancing* in a topless place"—the ambiguity of the studio versions evaporates. At another point the woman seems to be in a play of some sort, maybe based on Milton's *Paradise Lost*—but in a secular-looking establishment, with the rhythm of the final version audible in the words of the draft (9):

Called for you back stage that night but it was to {sic} easy
for you to leave
The 2nd act had just begun where Adam first meets Eve
I drifted into the audience of cattle dealers and pimps
Blue smoke rising from the
I tried to catch a glimpse

{illegible} (we) (you) I got too overly involved
Called for you backstage that night
I think you were in a trance
The Prince of Darkness blew his lines
So I thought I'd take a chance

The woman is also with someone else, a too-young rival nowhere visible in the song that would emerge from the notebooks: "That new boy hanging by your side, you'll teach him what {illegible, presumably "to do"?} / He must be all of 17, hey darling, shame on you." This is just a sampling of the different versions of the woman who is in the singer's past and perhaps, as the last stanza of the song allows, also in his future: "So now I'm going back again / Got to get to her somehow"—"gonna find her (get to her) somehow" in the notebook.

The third verse of the song already on various albums had a number of variants of the jobs the wandering singer has held but couldn't keep: in the great north woods working as a cook, drifting down to New Orleans, working on a fishing boat outside of Delacroix (also in Louisiana), loading cargo onto a truck, and so on. We can now add: "Used to work up in Oregon, with 20 men in a shack (helped build, logging) / Never did like the hours too much and one day I got the axe." Another variant is hilarious but would have to go, since it crossed the line into the openly autobiographical:

So I departed down to LA
Where I (reckoned) met my cousin Chuck
Who got me a job in an airplane plant
Loading cargo onto a truck.

As Dylan reveals in *Chronicles: Volume One,* his mother arranged the first accommodations for her eighteen-year-old freshman son when he arrived in Minneapolis in the summer of 1959:

> My mother had given me an address for a fraternity house on University Avenue. My cousin Chucky, whom I just slightly knew, had been the fraternity president . . . my mom said that she'd talked to my aunt about calling Chucky and letting me stay there.

As he crafted and recrafted these jobs in "Tangled Up in Blue," Dylan was clearly having fun, his mind roaming around the country, imagining the jobs that he himself never had in real life. As he sings with some self-irony on the 2006 song "Workingman's Blues #2," "Some people never worked a day in their life / Don't know what work even means." He knows what work means, but not jobs like these.

Another of the album's great songs, "Idiot Wind," in draft has lines wildly different from the versions that came out in

1975 or 1991, with words that are more revealing about the identity of Bob Dylan than he has anywhere let out (22):

People all have a different idea
Of who I am but I'm {illegible, probably "sure"} it's true
That none of them know what I'm really like
I don't know, maybe it's the same for you.

"Idiot Wind," rightly seen as among those songs that most directly confront what was going wrong with the marriage, shows Dylan struggling to catch the precise tone, which it would never finally succeed in doing—the reason for its success as art: "You're an idiot, babe," but also "We're idiots, babe," sharing the blame for what went wrong. There are four lines in the notebook of which there are no traces in the song:

I gave you {illegible} and soft summer rain
But you weren't contented 'til you saw me in pain
You took my blood babe, hope it gave ya a thrill
Now it's my turn, I'm gonna give up the bill

Apart from the enigmatic last line, these verses clearly go too far for the mix of anger and regret that make it onto the final versions. Whatever experience lies behind the song—and everyone knows what that is—the maturity of the songwriter jettisoned the lines along with a sentiment that no longer fits. The notebook is invaluable in showing us precisely the process Dylan the songwriter goes through as he struggles to reconcile experience and imagination in the interest of making a song.

Finally, for more than forty-two years, like countless others I have had in my head this version of the first verse on the album's song "Meet Me in the Morning":

Meet me in the morning, 56th and Wabasha
Meet me in the morning, 56th and Wabasha
We could be in Kansas
By the time the snow begins to thaw

In those days before Google Maps or any Internet at all, I had no idea where Wabasha might be, not that it really mattered. It in fact turned out to be in Minnesota, where it runs through downtown St. Paul, south across the Mississippi. There is no Fifty-Sixth Street in St. Paul, but Wabasha does intersect with Fifth Street and Sixth Street, a few blocks west of Highway 61 and eight miles east of Dinkytown, the place Bob Dylan spent those sixteen months honing his musical skills before heading to New York City in January 1961. On page 1 of the blue notebook there is a different beginning to the verse, with only the ending surviving the process of rewriting:

Meet me in the morning {illegible} we could have a ball
My grandfather had a farm but all he ever raised was the dead
He had the keys to the kingdom but all he ever opened was his head
Meet me in the morning, it's the brightest day you ever saw
We could be in Kansas by the time the snow begins to thaw

There are currently hundreds of books on Bob Dylan, the best of them clocking in at six hundred to nine hundred pages. We are only started on the long road that is the phenomenon of Dylanology. As Dylan sang on "Mississippi," "Stick with me baby, stick with me anyhow / Things could start to get interesting right about now."

8

MODERN TIMES AND THE WORLD'S ANCIENT LIGHT: BECOMING HOMER

I'VE BEEN CONJURING UP ALL THESE LONG DEAD SOULS FROM THEIR CRUMBLIN' TOMBS

—BOB DYLAN, "ROLLIN' AND TUMBLIN'"

The release of Bob Dylan's thirty-second studio album, *Modern Times,* on August 29, 2006, was attended by a sense of great anticipation from fans who were now reveling in Dylan's third "classic" period. Twice before in Dylan's career, peaks had descended to valleys—relatively speaking. *John Wesley Harding,* released in 1967, was and is a fine album, but it was not what those electrified by *Blonde on Blonde* from the year before had hoped for. Similarly, *Desire,* coming on the heels of the 1975 masterpiece *Blood on the Tracks,* is among Dylan's best albums, but fails to attain the heights of what had come before. Would *Modern Times* point back in the direction of *Under the Red Sky* from 1990, and other materials out of whose ashes—again, relatively speaking—*Time Out of Mind* and *"Love and Theft"* had risen? Or would what had been given back to Dylan—and to all of us—since 1997 keep on going? The latter proved to be the case, by near-universal assent. *Rolling Stone* pronounced the album Dylan's "third straight masterwork," and within two weeks *Modern Times* had topped the charts on the Billboard 200, the first Dylan album to do so in the thirty years that had passed since *Desire* hit that mark in 1976. *Modern Times* came out five years after *"Love and Theft"* and it would be another six before the 2012 masterpiece *Tempest*.

This might seem a very different pace than Dylan's period from 1964 to 1966, when all those songs came tumbling out. But in the early 2000s, Dylan was doing much more than just writing songs. He'd cowritten and starred in the 2003 movie *Masked and Anonymous,* a demanding, if hugely underrated, piece of work; in 2004 he'd published his bestselling memoir, *Chronicles: Volume One;* he was performing in concert at twice the rate that he had been in the sixties; he was painting and producing metal sculpture; and he recorded three seasons of his radio show *Theme Time Radio Hour* from May 2006 to April 2009. In 2008, moreover, he released *The Bootleg Series Vol. 8: Tell Tale Signs,* a trove of unreleased and variant versions of his songs from 1989 to 2006. In hindsight, the passage of time is trivial given the enduring achievement of what these years have produced: Dylan's music from the late twentieth and twenty-first century is a body of work that is comparable with that of any other period from Dylan's, or any artist's, career.

From the very beginning, the songs of *Modern Times* seemed rich in narrative texture, old and weary, mystical, musically varied, the songs alternately of a bluesman whose lyrics revealed the passing of time and of a highly poetic troubadour, still on the road after all these years. And these songs too take us back into another age. Even before the album came out, its title had piqued curiosity. Did the modern times suggest a connection with the 1997 masterpiece, *Time Out of Mind*? That album had reached back into the eighteenth and nineteenth centuries and beyond, to Robert Burns and the long traditions that went even further back in Burns's own folk heritage. And how did the 1936 Charlie Chaplin movie *Modern Times* enter into things? For those who know the movie, Chaplin's exploration of the desperate plight of the worker in the Great Depression resonates in the opening of one of the album's songs, "Workingman's Blues #2": "The buying power of the proletariat's gone down / Money's getting shallow and weak." Once 2008 rolled around, almost bringing a new Great Depression for the new century, the song's lines seemed more prophetic than retrospective.

In the song "When the Deal Goes Down," one of those great songs last played and as it seems laid to rest in Rome on November 7, 2013, Dylan expresses a sense of fatigue with "this earthly domain, full of disappointment and pain," and yet there is contentment and a readiness for what lies ahead, a resolution that whatever went wrong, "We learn to live and then we forgive / O'er the road we're bound to go." This particular song, like most of those on *Modern Times,* is overloaded with images, its narrative logic not entirely graspable. But it is a work of great beauty. In the lyrics there is a deep sense of humanity, and of survival. And how does the setting of the opening line, in "the world's ancient light," jibe with the modern times of the title? "When the Deal Goes Down" comes close at times to the equally powerful 1997 song "Not Dark Yet." Where that song gave us "I was born here and I'll die here against my will," this song ends its first verse, "We live and we die, we know not why / But I'll be with you when the deal goes down." "Not Dark Yet," in which the singer seems utterly alone, offered little in the way of hope, beyond the gift of the song itself. "When the Deal Goes Down" seems to have achieved some degree of hope, expressed through proximity to the addressee. The singer is not alone.

Director Bennett Miller's video of the song, starring Scarlett Johansson, captures its mood brilliantly, getting to the heart of the nostalgia and longing of Dylan's words. As the song opens, somewhere in time, "In the still of the night, in the world's ancient light," perhaps 1957, the video begins with a jerky, handheld 8mm filming of the central character in summer frock and sunglasses, a girl from the north country of Bob Zimmerman's youth. The then twenty-one-year-old actress could be a few years younger as someone films her on a boat in New York Harbor, the Statue of Liberty caught in the background. She is soon heading back home in a Chevy Bel Air convertible, its Minnesota plates revealing the year 1955. We are moved back in time, out of the world of the present, with its "earthly domain, full of disappointment and pain," to a time and a world whose passing Dylan has often noted. Dylan has moved back there in his song, for instance. The year could be 1958, the Minnesota girl could be Echo Star Helstrom, Bob Zimmerman's first steady girlfriend, and who knows, Bob could be the man behind the camera. Or not.

The girl is also at an aquarium, and on a lake in which she catches a fish. The town in the background could be Duluth. Three props make their cameo appearances, all sending us back in time to the three men he admired most: Woody Guthrie's *Bound for Glory* appears in the woman's lap as she swings in a porch hammock, a cassette of Hank Williams's *Wanderin' Around* is juxtaposed with an acoustic guitar propped up on the porch, and the video closes with what looks like a scene from Mardi Gras with masked figures, and the sleeve of the 1958 album *Buddy Holly*. The seventeen-year-old Dylan famously saw Buddy Holly at the Duluth National Guard Armory on January 31, 1959, three days before he died in the plane crash. Dylan has pretty much moved back there anyway, as in the new lyrics in performance of "Simple Twist of Fate": "She should have known me in '58 / She would have stayed with me."

Some songs on *Modern Times* were demonstrably old. Even

before we heard the words, once the title list was known, Dylan aficionados easily tracked down the eighth cut, "Nettie Moore," connecting it to an 1859 song, "Gentle Nettie Moore." There the singer is a slave whose woman has been bought by a trader from "Louisiana Bay" and taken off in shackles, leaving him alone in the little white cottage they had shared on the Santee River in South Carolina. All that is left is to wait for the day when he meets her in heaven, "up above the skies." Dylan takes over the opening of the chorus, "Oh I miss you, Nettie Moore, and my happiness is o'er," and the final line of his chorus gets the spirit of the song written two years before the outbreak of the Civil War: "The world has gone black before my eyes."

The rest of the song is pure Dylan invention, on the face of it an absurdist assortment of images that take the listener in all sorts of directions, incorporating fragments of other songs and texts, for instance quoting from Delta bluesman Robert Johnson's "Hellhound on My Trail": "Blues this morning falling down like hail." Dylan can juxtapose a reference to his own band ("I'm in a cowboy band"), to the excesses of Dylanology ("The world of research has gone berserk / Too much paperwork")—and then throw in a reference to the traditional folk song "Frankie and Albert," which he had covered on the 1992 anthology *Good As I Been To You:* "Albert's in the graveyard, Frankie's raising hell." And yet it works as a song whose sorrow reflects that of the 1859 slave song whose title it takes, but is intensified by the melody, the images, and above all by

Dylan's voice in all its aged richness. American historian and Dylan critic Sean Wilentz, who collects many of the components of the song, puts it well:

> The song wafts through time and space, past and present, old songs and new, as Dylan's recent songs do. It presents itself in the fragmented, ambiguous way that has marked Dylan's music, through many phases, for decades.

It is a world that can be comprehended only from inside the song, on its own terms. However, as always, recognition of the specifics of past and present, old and new are what can help inform and enrich the songs as they conjure up the way Dylan's thought shifted as he transforms his musical and literary traditions.

The sense of the old is conveyed not just by way of the intertexts of this album that inject their original contexts into the new setting, as with "Nettie Moore," but also by the very language that Dylan employs and by the settings he chooses: "I was passing by yon cruel and crystal fountain," "the whole world which people say is round" ("Ain't Talkin'"), "I'm staring out the window of an ancient town" ("Beyond the Horizon"). The week before *Modern Times* came out, Jon Pareles, journalist and music critic for the *New York Times,* well captured the essentials of the album:

> His lyrics, and sometimes his music, are studded with quotations and allusions spanning more than a century of Americana. . . . For Mr. Dylan there's no difference between an itinerant bluesman and a haggard pilgrim. "I practice a faith that's been long abandoned," he sings. "Ain't no altars on this long and lonesome road" . . . "The suffering is unending," he sings. "Every nook and cranny has its tears." He's a weary traveler, a bluesman and a pilgrim, on a dark and unforgiving road.

CONFEDERATE POET AND ROMAN EXILE

A few weeks after Pareles's review, on September 14, 2006, Motoko Rich reported in the *New York Times* that Albuquerque disc jockey Scott Warmuth had, through "judicious googling," revealed that a number of songs from *Modern Times* borrowed closely from Henry Timrod, a Confederate poet, born in Charleston, South Carolina, in 1828. Timrod wrote with a lyric voice that clearly appealed to Dylan, and the "old" feel of some of Dylan's lyrics seemed to have come straight from Timrod, as in the third verse of "When the Deal Goes Down":

> *More frailer than the flowers, these precious hours*
> *That keep us so tightly bound*
> *You come to my eyes like a vision from the skies*
> *And I'll be with you when the deal goes down.*

These lines, like a number of others on the album, draw heavily from two of Timrod's poems:

"A Rhapsody of a Southern Winter Night"

These happy stars, and yonder setting moon,
Have seen me speed, unreckoned and untasked,
A round of precious hours.
Oh! Here, where in the summer noon I basked,
And strove, with logic frailer than the flowers,
To justify a life of sensuous rest,
A question dear as home or heaven was asked,
And without language answered, I was blest.

and "A Vision of Poesy":

A strange far look would come into his eyes
As if he saw a vision in the skies

We are again reminded of the T. S. Eliot line: "Immature poets imitate; mature poets steal." The Bob Dylan of *Modern Times* is at the height of his maturity, and has here successfully stolen. The rhyming phrases "precious hours" and "logic frailer than the flowers" are unconnected except by rhyme in Timrod; Dylan brings them tightly together, transforming his source by having the hours, the shared time of the singer and the addressee, Dylan and us, be that which is frail or fragile, the internal rhyme "flowers . . . hours" matching that in his reuse of the other Timrod intertext, in a different poem, "eyes . . . skies," which is similarly transformed and bettered. Here Dylan has put himself back into his favorite century, and the song is all the more powerful once we see where this part of it came from and can appreciate how it was transformed. Again, look at Timrod, consider the two poets side by side, then ask yourself if you could do this:

More frailer than the flowers, these precious hours
That keep us so tightly bound
You come to my eyes like a vision from the skies
And I'll be with you when the deal goes down.

And that's quite apart from coming up with the melody and singing it with the cadence and voice of Bob Dylan. But *Modern Times* conjures up souls much longer dead than Timrod.

It turned out that Bob Dylan had again gone back to Rome in his songwriting. Soon after *Modern Times* came out in the fall of 2006, I was on sabbatical at Oxford University, hardwiring the new album, and getting lost in its beauty by listening to it every day on my walk back and forth to the library. I was working on a commentary on the last book of the *Odes* of the Roman lyric poet Horace (65–8 BC). As on Dylan's new album, some of Horace's poems connect song, love, and the passing of time, all filtered through the lyric genius of the Roman poet, for whom time is slipping away, with music and song a compensation, as at the close of the *Odes* 4.11, "To Phyllis":

> Love only as it is fitting; do not desire
> That which you ought not to have. Phyllis, listen to
> me:
> You are the last of my loves; there will be no others.
> Come, learn a new song and sing it to me, for song,
> Is the means, in your beautiful voice, to alleviate
> sorrow.
>
> —Horace, tr. David Ferry

With Horace's Phyllis as "the last of my loves," that particular poem had long since found a resonance in my mind with Dylan's 2001 song "Bye and Bye," where he had already sounded to me like Horace: "Well the future for me is already a thing of the past / You were my first love and you will be my last." This was a coincidence, but a meaningful one: the tradition Dylan is in goes back to the lyric poets of Greece and Rome.

So, there I was, in Oxford Town, working on the lyric poems of Horace and listening to the lyrics of Dylan, having taught courses on both of them a couple of years before, as I would keep on teaching them in the years that followed. When the album came out, I was also curious, given Dylan's quoting Virgil five years before on "Lonesome Day Blues," about what might be hiding out on the new album. What looked like one allusion was there in plain sight, on the first song, the driving "Thunder on the Mountain," in the sixth verse: "I've been sitting down studying the art of love / I think it will fit me like a glove." The only *Art of Love* I knew was the three-book poem of Roman poet Ovid (43 BC–c. AD 17), a playful early work, a "how-to" for those looking to get and to keep a romantic partner. But *Modern Times* didn't seem to have much to do with that poem, but rather, if anything, with the last poems Ovid wrote.

In AD 8, Ovid was exiled by Emperor Augustus—we really don't know why, possibly for writing the *Art of Love*—to a frontier Black Sea town, modern Constanta in Romania, about as bleak a spot as could be found for punishing the urbane, witty Roman poet. He spent the rest of his days there. He explored this exile in two collections of poetry, produced in the last years of his life: *Tristia* (*Poems of Sadness*) and *Letters from the Black Sea*. "Thunder on the Mountain" closed with the singer getting away from the world, heading north, another version of "Highlands" from *Time Out of Mind:* "I'm gonna make a lot of money, gonna go up north / I'll plant and I'll harvest what the earth brings forth." This ending to the first song seemed, along with the final verse of the last song, to frame the whole album:

Ain't talkin', just walkin'
Up the road around bend
Heart burnin', still yearnin'
In the last outback, at the world's end

That ending was also stolen, from Ovid. On October 10, 2006, Cliff Fell, a New Zealand poet and teacher of creative writing, wrote in his local paper, the *Nelson Mail,* of a striking discovery. He happened to be reading Peter Green's Penguin translation of Ovid's exile poetry. Green is one of the finest translators of Greek and Latin poetry, always coming up with the right idiom for his authors, and so bringing them to life in contemporary English. This is an important detail. Like me in Oxford and like millions throughout the world, Fell was also listening to *Modern Times*. As he describes it:

> and then this uncanny thing happened—it was like I was suddenly reading with my ears. I heard this line from the song "Workingman's Blues #2," "No-one can ever claim / That I took up arms against you." But there it was singing on the page, from Book 2.52 of *Tristia:* "My cause is better: no-one can claim that I ever took up arms against you."

As Fell read on in Ovid, he came across further lines that were entering his consciousness from listening to *Modern Times*:

Bob Dylan, "Ain't Talkin'"	**Ovid, *Black Sea Letters*, 2.7.66**
Heart burnin', still yearnin' In the last outback at the world's end.	I'm in the last outback, at the world's end.
Bob Dylan, "Workingman's Blues #2"	**Ovid, *Tristia*, 5.12.8**
To lead me off in a cheerful dance.	or Niobe, bereaved, lead off some cheerful dance.
Bob Dylan, "Workingman's Blues #2	**Ovid, *Tristia*, 5.13.18**
Tell me now, am I wrong in thinking That you have forgotten me?	May the gods grant . . . / that I'm wrong in thinking you've forgotten me!
Bob Dylan, "Workingman's Blues #2"	**Ovid, *Tristia*, 2.179**
My cruel weapons have been put on the shelf / Come sit down on my knee /	Show mercy, I beg you, shelve your cruel weapons.
Bob Dylan, "Workingman's Blues #2"	**Ovid, *Tristia* 5.14.2**
You are dearer to me than myself / As you yourself can see.	wife, dearer to me than myself, you yourself can see.

The recognition that comes from reading or hearing one text through the medium of a later text is part of the aesthetic pleasure that is the product of the intertextual process, and the excitement of Fell's discovery was apparent.

It eventually emerged that more than thirty lines of Ovid's exile poems had been reappropriated and become an essential part of the fabric of the songs of *Modern Times*. Dylan even reused Ovid in the title of the first song of the next album, *Together Through Life,* from 2009. "Beyond Here Lies Nothin'," a song firmly in the setlist of concerts, is lifted from Ovid, *Tristia* 2.195–96: "beyond here lies nothing but chilliness, hostility, frozen / waves of an ice-hard sea." The refrain of the third verse seems to look back to the source: "Beyond here lies nothin' / But the mountains of the past."

Ovid was one of those "mountains of the past." Scaling the "mountains of the past" was something Dylan continued to do in his covers of the Great American Songbook in 2015–17, with Frank Sinatra now replacing Ovid. So, from an interview coinciding with release of *Shadows in the Night:*

> You know, when you start doing these songs, Frank's got to be on your mind. Because he is the mountain. That's the mountain you have to climb even if you only get part of the way there.

Scaling the mountains of the past is expressed in a different metaphor in the lively blues song "Rollin' and Tumblin'": "I've been conjuring up all these long dead souls from their crumblin' tombs." That line too is borrowed from Ovid, not his exile poetry, but rather the love poems of his youth—where a witch "conjures up long-dead souls from their crumbling sepulchers." Timrod and Ovid are the long-dead souls, joining those of Homer, Virgil, Burns, and any number of others whose poetry and song Dylan has been putting to such good use in these years. Bob Dylan has been conjuring them up, as he brings them back to life in the songs of *Modern Times*, in effect bringing *them* into the modern times, where they fit so well, both in translation and in their transformations into Dylan's songs.

This connects to other conjuring Dylan has been doing of late. Toward the end of the interview Dylan did with Mikal Gilmore for the September 27, 2012, issue of *Rolling Stone,* Dylan was asked about his "quotes" of Timrod:

> And as far as Henry Timrod is concerned, have you ever heard of him? Who's been reading him lately? And who's pushed him to the forefront? Who's been making you read him?

The answer to these slightly elusive questions is "Bob Dylan." It is Dylan who has brought the obscure Timrod and the less obscure Ovid out from their tombs into the full light of day. Dylan has done this because he cares about the traditions in which he belongs, about poetry and music, and about taking us back into those lost worlds that are so vital a part of him. Incidentally, in his very response to Gilmore, we hear a line from "Soon After Midnight," a song on *Tempest,* released two weeks before the interview: "Two Timing Slim, who's ever heard of him."

WHY THE EXILE POEMS?

In the inner exile he created for his own protection, and from which he sends us his songs, Bob Dylan discovered and invoked Ovidian exile poetry, the poetry coming at the end of the career of Ovid. Indeed, the last words of the last song, "Ain't Talkin'," and therefore the last words of the third album of what looked like the trilogy, suggest a finality, a closing of the book, and they are straight from Ovid as Dylan puts himself "in the last outback, at the world's end." At the same time, it needs to be noted that Dylan's borrowings—or thefts—are all transposed into new situations that have little to do with, but that once noticed and activated evoke comparison with, those of the Ovidian models—the essence of creative intertextuality. It is worth noting again the perception of T. S. Eliot, who himself practiced precisely the art that Dylan had become part of:

> The good poet welds his theft into a whole of feeling which is unique, utterly different from that from which it is torn. . . . A good poet will usually borrow from

authors remote in time, or alien in language, or diverse in interest.

That is part of the art of Bob Dylan. On songs like "Ain't Talkin'" and "Workingman's Blues #2" he is not citing or quoting; rather he is renewing and re-creating, as he has been doing for years, with material from folk and other traditions, but also the Bible, Rimbaud, and more.

The Ovidian lines reused by Dylan are largely acknowledged now. But why Ovid? What is it about the Roman poet that made his voice, traveling across two thousand years into Green's translation, so appealing to Dylan? Ovid is very different from Cicero, who was also exiled and also wrote real letters back to friends, his wife, his brother, and various other figures. What Ovid wrote was exile *poetry,* many of the poems posing as letters, but they are not actual letters. The poems lament his condition, but they still show the wit, irony, and character of the poet familiar to readers of Ovid's earlier and happier times, for instance the reader of the *Metamorphoses,* one of the texts encountered by Dylan in Ray Gooch's library in *Chronicles: Volume One*. Ovid's exiled voice is not just that of a sufferer; he is also in control of things, doing with his lyrics just what he wants, creating the persona, or exile mask, he wants for any given poem. All we have of Ovid is his poetry, what he wanted the world to see of him, and in this respect he and Dylan belong together.

There is no evidence outside his own poetry that Ovid ever went into exile, which is strange given the prominence of the poet. We might have expected some later historian to mention the exile. There is in fact a view, far from an orthodoxy, but not in my view beyond belief, that Ovid never went into exile, on the Black Sea or anywhere else. In this line of thinking he could have been living in Rome, a villa on the Tiber perhaps, heading in the summer for the trendy Bay of Naples or the cool of his native Sulmona, in the mountains ninety miles north of Rome. This possibility does not need to be true in order to see Ovid's exile poems for what they are, poetic constructions, practicing the essence of art that he—and Dylan—knew long before Rimbaud said it: "I is an *other*." Ovid's exile poems are exercises in the genre of exile poetry, artistic creations of the voice of one suffering from solitude in a hostile, unwelcoming setting at the ends of the earth. That is how and why Dylan was attracted to them as he created the masks and voices of the songs on *Modern Times* that look back to the Roman poet.

MEMORY, SONG, AND NOSTALGIA

Ovid's poems are also powerful in the nostalgia they evoke, a nostalgia for the city he has lost, in reality or in his imagination, nostalgia also for absent friends and family. The Greek root means pain, *algos* for return home, *nostos*. The *Odyssey* was one of a number of poems about the return home, or failure to return home, of the Greek leaders after they sacked Troy. As a group these poems were called *Nostoi*. In one of the affectionate poems addressed to his wife, Ovid directed his solitary song to the stars and the night sky:

> Turn your glistening faces on my lady, and tell me
> Whether she thinks of me or not. Alas,
> why seek the answer to what's only too apparent?
>   Why do
> my hopes slide into fear and doubt? Believe
> what's as you would want, quit agonizing over
> what's secure: bet safe on a safe bet,
> and what the gleaming pole stars cannot tell you,
> now tell yourself in veridic utterance:
> *That she, who's your prime concern, has never forgotten*
> *Your memory, that she cherishes your name*
> *(all that's left her of you), dwells ever on your features*
> *as though you were present; and though far away,*
> *if you still love, still loves you*
>
> —Ovid, *Tristia* 4.3, tr. Green

Clearly this nostalgia and play on memory and forgetting, absence and uncertainty, is what gives Ovid's poems the very universal human appeal that they have for readers, Dylan included. Two examples from "Workingman's Blues # 2" both come straight from Ovid, who at *Black Sea Letters* 4.6.42–43 has

"*Them* I'll forget, / but *you* I'll remember always," contrasting his friend Brutus with other friends who have betrayed him. But in both cases Ovid is only the jumping-off point for something more intense and developed. So Dylan gives us "Them, I will forget / You, I'll remember always." The verse, as sung on the album, then continues with a reiteration of the nostalgia and remembering: "Old memories of you to me have clung / You've wounded me with words," then ends enigmatically, suggesting trouble, unspecified and mysterious, between the speaker and his addressee: "Gonna have to straighten out your tongue / It's all true, everything you have heard."

The second borrowing, or theft, of Dylan picks up on Ovid's anxiety about the possibility of his wife's forgetting him, since he has had no letter from her:

> May the gods grant . . .
> that I'm wrong in thinking you've forgotten me!
>
> —*Tristia* 5.13.18

Dylan converts this into a nostalgic reflection as the sun goes down and he wishes someone from the past were with him to see:

> *Tell me now, am I wrong in thinking*
> *That you have forgotten me?*
>
> —"Workingman's Blues #2"

When Dylan reworks Ovid's line in "Tell me now, am I wrong in thinking / That you have forgotten me?" what we have is what we have always had in Dylan, the world gone wrong for the lover. "Something's out of whack," as he sings on "Nettie Moore." That quality was what he discovered in Ovid, but it has been a feature of his songwriting from the beginning. Much of his genius has been to capture the pain of separation in space and time. That is the essence of folk music, and of the blues, in which memory of the past is what helps create song. From "Boots of Spanish Leather" back in 1963 and in many songs since, thoughts of the absent lover and the fear of being forgotten, or the pain that accompanies memory of someone or something, person, or place, is what goes into making the song so timeless. There is a reality behind that song, the well-known fact that Bob Dylan wrote it when he was separated from his girlfriend Suze Rotolo.

She was the Muse who gave him that song, but the *song* that Dylan created from that reality is the same even if we know nothing of the circumstances of its composition, even if the reality did not exist. The song alternates its first six verses between the singer who is left behind and the woman who is "sailing away in the morning" and asks what she can send him "from across the sea"—from Spain, it turns out, not Italy, where Suze Rotolo had gone in the summer of 1962. His request evokes the response that no gift can compensate for her absence: "Just carry yourself to me unspoiled / From across that lonesome ocean," the adjective "lonesome" transferred from him to the ocean that separates them. The word comes back again in his voice later in the song: "I got a letter on a lonesome day." He settles in the end for something she can send back, "Spanish boots of Spanish leather," a resolution of sorts. Her last verse comes in the middle of the song—he sings the last four—as she asks the question for the last time:

That I might be gone a long time
And it's only that I'm askin'
Is there something I can send you to remember me by
To make your time more easy passin'

His response then gets to the heart of the matter, the lover's nostalgia, for which there is no sufficient recompense:

Oh, how can, how can you ask me again
It only brings me sorrow
The same thing I want from you today
I would want again tomorrow

The beauty of this song came cross when he sang it, still a twenty-one-year-old, at the Town Hall concert in New York on April 12, 1963, as on November 6, 2013, when he seems to have bid it farewell in performance at Rome, in the country where he wrote it.

The other song Dylan partially wrote in Italy in 1962 or 1963 was "Girl of the North Country," with its framing third line in the first and last verses: "Remember me to one who lives there," followed by a reaching for the past that is gone but not forgotten, "She once was a true love of mine," along with the juxtaposed lines ending the third and beginning the fourth verse: "That's the way I remember her best / I'm a wonderin' if she remembers me at all." The melody and much of the beauty of the song come from its evoking of "Scarborough Fair," the British folk song Dylan had heard in London at the end of 1962, but the lyrics are all Dylan and are what bring this song to life. Similar evocations of the past run throughout the song list of Dylan:

"Idiot Wind," 1975
I can't remember your face anymore, your mouth has changed,
your eyes don't look into mine.

But then his memory returns:
I followed you beneath the stars, hounded by your memory
And all your ragin' glory.

"Isis," 1976
I still can't remember all the best things she said

and:
I still can remember the way that you smiled.

In "I'll Remember You," 1985, each of the three verses opens and closes with the title line.

"Most of the Time," 1989
Don't even remember what her lips felt like on mine.

"Tryin' to Get to Heaven," 1997
Every day your memory grows dimmer
It doesn't haunt me like it did before

"Til I Fell in Love with You," 1997
When I'm gone you will remember my name. . . .
I'm thinking about that girl who won't be back no more.

"Cold Irons Bound," 1997
I'm gonna remember forever the joy that we shared.

"Workingman's Blues #2," 2006
The place I love best is a sweet memory

"My Wife's Home Town," 2009
Well there's plenty to remember, plenty to forget
I still can remember the day we met

And finally, Dylan comes out and sings it as the leitmotif becomes the actual theme in *Together Through Life*'s "Forgetful Heart" from 2009, cowritten with Robert Hunter—a song frequently performed until Dylan replaced it with some of the American Songbook numbers that it so resembles and in hindsight was pointing toward:

Forgetful heart
Lost your power of recall
Every little detail
You don't remember at all
The times we knew
Who would remember better then you. . . .

Any number of examples could be added: "If You See Her, Say Hello," "Sara," "Shooting Star," "Red River Shore," all songs about the pain of remembering, songs whose beauty attracts us into the lyrics as we share in the aesthetics of remembering. Dylan did not need Ovid in order to bring out the nostalgia of remembering; indeed most of these instances above are from songs written before he read Green's 1994 translation. But in Ovid he found a kindred spirit, and that is why he took on the persona of the Roman poet precisely by integrating him into his own verse.

OVID BECOMES ODYSSEUS

Like Dylan, Ovid was a trickster, and he was also attracted to Odysseus ("Ulysses" for the Romans), the ultimate trickster and lyre-playing teller of tales true and tall: "Sing to me, Muse, of a man full of many twists and turns," as the *Odyssey* begins, a fine description of Odysseus, Ovid, and Bob Dylan. The exiled Ovid humorously compared himself to Odysseus in the exile poems that Dylan reused in *Modern Times:*

> He wandered for years, but only
> on the short haul between Ithaca and Troy
> He had his loyal companions
> His faithful crew; *my* comrades deserted me
> At the time of my banishment. He was driven from his
> homeland,
> A cheerful victor: I was driven from mine—
> Fugitive, exile, victim. My home was not some Greek
> island,
> Ithaca, Samos—to leave *them* is no great loss—
> But the City that from its seven hills scans the world's
> orbit,
> Rome, centre of empire, seat of the gods.
> I was crushed by a god, with no help in my troubles
> *He* had that warrior-goddess [Athena] at his side.
> What's more the bulk of his troubles are fictitious,
> Whereas mine remain anything but myth.
>
> —*Tristia* 1.5.59–80

Really? This last line could be truth or untruth, with either being well and good. By the end of the fourth book of these poems, Ovid takes the comparison even further and has more or less become Odysseus, as he relives the journey:

> By sea and land I suffered as many misfortunes
> As the stars between the unseen and the visible poles.
> Through long wanderings driven, I at length made
> landfall
> On this coast where native bowmen roam; and here,
> Though the din of neighbouring arms surrounds me, I
> still lighten
> My sad fate as best I can
> with the composition of verse: though there is none to
> listen
> this is how I spend, and beguile, my days.
>
> —*Tristia* 4.10.107–14.

Some may hear in the final line the last verse of "Pay in Blood," the 2012 song from *Tempest:* "This is how I spend my days / I came to bury, not to praise"—with the clear addition from another voice in Dylan's tradition, in Shakespeare's *Julius Caesar:* "I come to bury Caesar not to praise him." But moving from *Modern Times* to *Tempest,* we also move on to another development.

DYLAN BECOMES ODYSSEUS

> "[N]o one can touch you, much less beat your distance!"
>
> —Athena to Odysseus

Ovid "transfigured" himself into Odysseus, as Dylan would say, and Dylan, who had taken on the voice of Ovid on *Modern Times,* then followed suit and on *Tempest* got himself back to where it all began—with an intertextual brilliance that shows exactly how he sees his art and how to conjure up the long-dead souls of the poets he has been reading. Sometime around 2010, Dylan in performance abandoned the lyrics of the second-to-last verse of "Workingman's Blues #2," the one in which his voice borrowed from Ovid:

Dylan	**Ovid**
In you, my friend, I find no blame You wanna look in my eyes, please do No one can ever claim That I took up arms against you	No one can claim that I ever took up arms against you

In the performances of the last few years, as in the official text of *Bob Dylan: The Lyrics: 1961–2012,* the verse is gone, completely rewritten:

I'll be back home in a month or two
When the frost is on the vine
I'll punch my spear right straight through
Half-ways down your spine

I'll lift up my arms to the starry skies
And pray the fugitive's prayer
I'm guessing tomorrow the sun will rise
I hope the final judgment's fair.

No sign of Ovid, except perhaps in the "fugitive's prayer." Instead Dylan, perhaps led there by Ovid, had gone back to the *Odyssey,* with whose hero, "the man of twists and turns," he has long associated, specifically to Robert Fagles's 1996 Penguin translation of Homer's poem. Dylanologist Scott Warmuth gathered some of the Homeric intertexts, but what needs explaining is what these quotes are doing in Dylan's songs. Dylan transfigured is here quoting and channeling Odysseus, from Book 10 of the *Odyssey,* on the island of the witch and temptress Circe—"you too have shared a bed with the wrong woman," says Dylan in the Nobel lecture in June 2017 as he comes out and compares himself to Odysseus. The Greek hero is telling his host, King Alcinous, of going out to reconnoiter and killing a stag, dinner for his hungry crew, that some god sent his way:

Just bounding out of the timber when I hit him
Square in the backbone, halfway down the spine
And my bronze spear went punching clean
through—
He dropped in the dust, groaning, gasping out his
breath

In the second-to-last verse of the great 2012 song "Early Roman Kings," the Roman kings undergo one of a number of transformations, and things get very strange:

> ***I'll strip you of life, strip you of breath***
> ***Ship you down to the house of death***
> *One day you will ask for me*
> ***There'll be no one else that you'll want to see***
> *Bring down my fiddle, tune up my strings*
> *Gonna break it wide open like the early Roman kings.*

The singer utters the taunt the Greek hero hurls at the Cyclops he has just blinded in Book 9 of the *Odyssey:* "Here was my parting shot," Odysseus tells Alcinous:

> **Would to god I could strip you**
> **of life and breath and ship you down to the House**
> **of Death.**

As always when he means the intertextuality to be noticed, there is no question about its source. The line that Dylan adds—"There'll be **no one** else that you'll want to see"—cleverly represents Dylan's own free expansion of the Homeric situation. By slipping in the words "no one," Dylan points to the name Odysseus had given to the Cyclops, one of his many untruths.

Nor is the singer's identification with the hero limited to

this one song. In Book 5 of the *Odyssey,* the ship of Odysseus is wrecked by the storm that Poseidon (Neptune) sends in revenge for Odysseus's blinding of the Cyclops. This storm is the ultimate source of all the storm scenes in the Western literary tradition, including the one that opens Shakespeare's play *The Tempest*. Asked if the 2012 album was named after that play, Dylan replied in the negative: his was called *Tempest,* not *The Tempest,* a different title.

Odysseus, with the help of his patron goddess Pallas Athena ("Minerva" for the Romans), survives the storm, arrives at the island of the Phaeacians, and is challenged by one of the local princes to compete in the games the king is holding in his honor: "You're no athlete. I see that!" Odysseus silences the younger men with a discus throw that far surpasses their efforts, but only after delivering a speech of admonition to him. The speech obviously appealed to Dylan, an older man silencing his younger critics by demonstrating his superior strength. He borrowed five lines from the speech, scattering them across three songs of *Tempest:*

"You, you're a reckless fool, I see *that*"	*Odyssey* 8.192
"You're a reckless fool, I can see it in your eyes"	"Tin Angel"
"A god can crown his words with beauty, charm"	*Odyssey* 8.196
"but there's not a bit of grace to crown his words"	*Odyssey* 8.202

"She has crowned my soul with grace"	"Narrow Way"
"Just like you, my fine, handsome friend"	*Odyssey* 8.203
"Just like you, my handsome friend"	"Pay in Blood"
"But the mind inside is worthless"	*Odyssey* 8.205
"He's a gutless ape with a worthless mind"	"Tin Angel"
"despite so many blows"	*Odyssey* 8.213
"How I've survived so many blows"	"Pay in Blood"

This is not a matter of laziness or plagiarism. All of these references come from a speech of the mature Odysseus, whose identity the characters in Dylan's songs, and Dylan himself, are taking upon themselves. And when he sings these songs in performance, Dylan has had a classical statue next to him onstage. It is a river goddess, a likeness of a statue group of Pallas Athena (Minerva). The group is outside the Parliament building in Vienna—suggesting a connection between democratic Austria and the ideal Athenian democracy, of which Athena is the patron goddess. The same river goddess is on the cover of *Tempest*. Why? I would say because Dylan transfigured into Odysseus, the wandering survivor of so many blows, quite naturally has with him a statue associated with the goddess, has taken her on as his divine patron.

When Odysseus wins the contest, she exclaims that "no one can touch you, much less beat your distance!" Dylan has long known that the same applies to him. In early June 2010, when he was slipping the lines from Homer into performances of "Workingman's Blues #2" and writing the songs for *Tempest* in which his voice became that of Odysseus, he was touring in the Balkans and Eastern Europe. From June 9 to June 12, he circled Vienna, performing three concerts in three countries: in Bratislava, thirty miles east of Vienna across the Slovak Republic border, then north to Prague in the Czech Republic, and then back into Austria, at Linz, to the west of Vienna. My guess is that he dropped in on the Austrian capital, where he has performed a number of times. That's when it all came together: Homer, the Athena statue group, and the creative mind of Bob Dylan, who notices everything around him.

On March 22, 2017, just before the release of *Triplicate,* Dylan had a conversation with author Bill Flanagan exclusive to the official website bobdylan.com. Flanagan is too seasoned an interviewer of Dylan to have asked the old chestnut of a question, "What do you think of Joan Baez?" But that is what he asked, and Dylan's response may be heard as a commentary on the process here traced, Bob Dylan as Odysseus one more time:

> She was something else, almost too much to take.
> Her voice was like that of a siren from off some Greek
> island. Just the sound of it could put you in a spell.
> She was an enchantress. You'd have to get yourself
> strapped to the mast like Odysseus and plug up your

> ears so you wouldn't hear her. She'd make you forget
> who you were.

In Book 12 of the *Odyssey,* the hero ties himself to the mast and gets his crew to fill their ears with beeswax, so *they* cannot hear the enchanting song of the Sirens, which he listens to. Dylan is clearly having fun as he imagines himself in the role of Odysseus.

And once more, in his June 2017 Nobel lecture, with which this book will more or less conclude, Dylan expresses this connection between Odysseus the arch-trickster and himself:

> He's always being warned of things to come. Touching things he's told not to. There's two roads to take, and they're both bad. Both hazardous. On one you could drown and on the other you could starve. He goes into the narrow straits with foaming whirlpools that swallow him. Meets six-headed monsters with sharp fangs. Thunderbolts strike at him. Overhanging branches that he makes a leap to reach for to save himself from a raging river. Goddesses and gods protect him, but some others want to kill him. He changes identities. He's exhausted. He falls asleep, and he's woken up by the sound of laughter. He tells his story to strangers. He's been gone twenty years. He was carried off somewhere and left there. Drugs have

> been dropped into his wine. It's been a hard road to travel.
>
> In a lot of ways, some of these same things have happened to you. You too have had drugs dropped into your wine. You too have shared a bed with the wrong woman. You too have been spellbound by magical voices, sweet voices with strange melodies. You too have come so far and have been so far blown back. And you've had close calls as well. You have angered people you should not have. And you too have rambled this country all around. And you've also felt that ill wind, the one that blows you no good. And that's still not all of it.

Here "he" in the first paragraph is Odysseus, and "you" in the second paragraph is Bob Dylan, confirming his transfiguration, and listing a few of their shared experiences: some Circe of the sixties putting drugs in his wine; sharing the bed of his own Calypso, the "wrong woman"—whoever that might be in reality; angering those fans not ready for the changes he went through. But the "you" of this paragraph is not just Bob Dylan. It is also you and me, those who have been part of his odyssey. As he sang in "Mississippi," perhaps thinking of Odysseus as early as 1997, "I've got nothin' but affection for all those who've sailed with me." In this Dylan is not just channeling Homer. He seems to have added another poet, the great Constantine Cavafy, whose poem "Ithaca" expresses the idea that we all have, or should have, a little Odysseus—indeed a little Bob Dylan—in us. Here is the first verse of Cavafy's poem:

> As you set out toward Ithaca,
> hope the way is long,
> full of reversals, full of knowing.
> Laistrygonians and Cyclops,
> angry Poseidon you should not fear,
> never will you find such things on your way
> if your thought stays lofty, if refined
> emotion touches your spirit and your body.
> Laistrygonians and Cyclops,
> savage Poseidon you will not meet,
> if you do not carry them with you in your soul,
> if your soul does not raise them up before you . . .
>
> —Cavafy, "Ithaca," tr. Theoharis Theoharis

Dylan returns to the third person "he" as he closes his lecture on the *Odyssey*, with a focus on the trickster hero's homecoming and his dealing with the suitors who have been trying to woo his wife Penelope during his twenty-year absence, and who will pay with their blood for their misdeeds:

> When he gets back home, things aren't any better.
> Scoundrels have moved in and are taking advantage

> of his wife's hospitality. And there's too many of 'em. And though he's greater than them all and the best at everything—best carpenter, best hunter, best expert on animals, best seaman—his courage won't save him, but his trickery will.
>
> All these stragglers will have to pay for desecrating his palace. He'll disguise himself as a filthy beggar, and a lowly servant kicks him down the steps with arrogance and stupidity. The servant's arrogance revolts him, but he controls his anger. He's one against a hundred, but they'll all fall, even the strongest. He was nobody. And when it's all said and done, when he's home at last, he sits with his wife, and he tells her the stories.

Though ostensibly speaking of Odysseus, with an allusion to the Cyclops scene ("he was nobody"), Dylan is again pointing to his own song, and chiefly to "Pay in Blood," one of the most *Odyssean* songs on *Tempest*, and one that he sings night after night, with his Minerva statue behind him, backing him up:

> How I made it back home nobody knows
> Or how I survived so many blows
> I been through hell, what good did it do?
> My conscience is clear, what about you?

Again "Nobody," aka Odysseus, knows how the singer found his homecoming, like Odysseus the singer has survived blows and been through hell, or the underworld, and as the repeated refrain of the next verse affirms: "I pay in blood but not my own."

This method of composition is not to be thought of as mere quotation or citation. Rather it is a creative act involving the "transfiguring" of song and of literature and of characters going back through Rome to Homer. It is the means by which Dylan imagines and creates the worlds that he then inhabits in his songs and in performance. He has been up to this mostly in the years since *"Love and Theft"* in 2001 but his songwriting has always come from other places, drawing its meaning from them, not least the worlds of the Greeks and Romans. In the process Dylan becomes part of the stream that flows from Homer on into the present. As he said at the close of his Nobel lecture: "I return once again to Homer, who says, 'Sing in me, o Muse, and through me tell the story.'" Long may that story run.

9

THE SHOW'S THE THING: DYLAN IN PERFORMANCE

Like the poems of the Greeks and Romans, Bob Dylan's song is meant to be experienced in performance. Dylan himself captured the essence of the matter in an interview with music critic Jon Pareles, two days before the release of his 1997 comeback album, *Time Out of Mind:*

> A lot of people don't like the road, but it's as natural to me as breathing. . . . I'm mortified to be on the stage, but then again, it's the only place where I'm happy. It's the only place you can be who you want to be.

In the words of Bob Dylan, "Any minute of the day, the bubble could burst" ("Sugar Baby," 2012). When that day comes, the Tulsa archive and other resources will preserve a simulacrum, an image or likeness, of the man and his performance. As such, it will be without full human essence, and it is that yearning for that human experience that keeps us coming back to Dylan.

That's why I decided to get myself down for one of the last shows of the 2016 fall tour, in Clearwater, Florida, on November 19. "Had to go to Florida," as the 2001 song "Po' Boy" put it. All 2,180 seats at Ruth Eckerd Hall were filled, a beautiful venue with perfect acoustics on the shores of Alligator Lake, south of Safety Harbor. I had last seen Dylan and his band earlier in the year, on July 14, that time on the Boston waterfront. I partly felt an urge to see a post-Nobel Dylan concert. I was pretty sure nothing would be revealed in Clearwater or anywhere else—just a performance, from Stu Kimball's opening guitar stage left, to Dylan and the band lined up and motionless under the closing lights. These days you have the words, the song, the band, and you have voice, gesture, and presence of Dylan, and that's likely all there will be from here on out. The concert was brilliant from start to finish, as revealed on a recording available on YouTube at the time of writing. There was also something about the story or drama that Dylan's setlist had become that induced me to get one last concert in. I had gotten an inkling in Boston, and it came home powerfully in Clearwater.

UNCOVERING THE GREAT AMERICAN SONGBOOK

Six of the songs in the concert were from the new "cover" albums, the only studio recordings that Dylan has put out since

Tempest in 2012. In Boston the number of those songs had been eight, including the final performance of the tour for Irving Berlin's song from 1932, "How Deep Is the Ocean (How High Is the Sky)?" David Kemper, a drummer for the Jerry Garcia Band who played with Dylan from 1996 to 2001, tells of going into the studio for four days before a tour, also doing old songs, like Dean Martin's "Everybody Loves Somebody Some Time." When the band recorded *"Love and Theft"* in 2001, Kemper recalls Dylan saying:

> "All right, the first song we're going to start with is this song," and he'd play it on the guitar and then he'd say "I want to do it in the style of this song," and he'd play an early song. Like he started with "Summer Days" and he'd play a song called "Rebecca" by Pete Johnson and Big Joe Turner. . . . It was like, "Oh my God, he's been teaching us this music [all along]—not literally these songs, but these *styles*."

Some years later, Dylan would begin to bring such songs into his concerts, at first just one song. On October 26, 2014, he was in Hollywood for the last of a three-night stand at the Dolby Theatre. He and his band closed with a single encore, "Stay With Me," a song written by Jerome Moross and Carolyn Leigh for the 1963 film *The Cardinal,* starring Tom Tryon. It is a short song, at home in that movie, where it comes across as a sort of prayer:

Should my heart not be humble, should my eyes fail to see,
Should my feet sometimes stumble on the way, stay with me. . . .

Once Frank Sinatra covered it in 1964, the song became more secular, though no less poignant, without any specific cultural context. By the end of the fall of 2014, in the hands of Bob Dylan, "Stay With Me" had, in the words of music journalist David Fricke, been turned into "the most fundamental of Great American Songs: a blues."

The success in performance of "Stay With Me" may have helped Dylan decide to lay down these American standard songs of the middle third of the twentieth century—"the same songs that rock 'n' roll came to destroy," as Dylan put it in an interview with Robert Love. *Shadows in the Night* was released on February 3, 2015. Dylan was clearly proud to have his five-piece band backing what was initially described as a "Sinatra cover album" but came to be seen as something more than that as the other albums followed. In place of an orchestra, Dylan's voice is accompanied by the touring band that has now been with him for quite some time, with horns added for some songs. The critical response was overwhelmingly positive, with *Triplicate* topping the British charts within a week of its release. Once again, Dylan knew what he was doing, and what he was doing was "uncovering," not covering, these songs, bringing them back to life as surely as he had brought Homer, Virgil, Ovid, and Timrod back from their "crumblin' tombs."

In the interview with Love, Dylan confirmed Kemper's memory of playing such songs as far back as 2001, when the drummer left the band. The songs, Dylan revealed, all seem connected one way or another: "We were playing a lot of these songs at sound checks on stages around the world without a vocal mic, and you could hear everything. You usually hear these songs with a full-out orchestra. But I was playing them with a five-piece band and didn't miss the orchestra."

For these songs, the studio was only the start. They had started out in sound checks, Dylan's way of teaching his band about the old "standards," and had then been taken into the studio. Now they were back onstage, in performance, almost a third of the setlist in recent performances. That's where they needed to be heard to see how well they fit particularly with Dylan's own more recent songs. With the exception of "Some Enchanted Evening," all the songs from *Shadows in the Night* soon joined "Stay With Me" and became part of Dylan's performance, through the spring and summer concerts of 2015.

In reality, Dylan has given new life to these songs, particularly in performance. They are not just revived, they are transformed, even transfigured, by virtue of Dylan's incorporating them into his own story. If you listen to the forty-one-year-old Sinatra singing "Autumn Leaves" in 1957, backed by an orchestra and in his full maturity, that is a fine experience, but it is just a song, covered by Sinatra, sung beautifully. Hearing Dylan end a concert with his version of the song, with Donnie Herron's steel guitar lead-in replacing the string section—and outdoing the orchestra in its plaintive qualities as an interpretation of the song—is a different experience, because Dylan, aged seventy-five, also singing beautifully, had integrated the song into the story of his own songbook.

On December 20, 2016, NBC aired *The Best Is Yet to Come,* a concert in honor of the singer Tony Bennett's ninetieth birthday a few days earlier. The show was filmed on September 15 at Radio City Music Hall. On October 28, on the band's day off between concerts at Jackson, Mississippi, and Huntsville, Alabama, Dylan went into Workplay Studios in Birmingham, Alabama, and recorded a video of a Charles Strouse and Lee Richard Adams song from 1962, "Once Upon a Time"—covered by Bennett, Sinatra, and others—which was then shown during the NBC concert. It would eventually be released on the first disc of *Triplicate*. Both there and when I first heard it on TV, it somehow took me back to the great 1965 song "Like a Rolling Stone," which shares its opening with the title of the "cover song," "Once upon a time. . . ." In that studio Dylan's words and the band's accompaniment express the melancholic sense of a world that is past, never to be brought back: "Once upon a time, the world was sweeter than we knew / Everything was ours; how happy we were then / But somehow once upon a time never comes again." If that is too melancholy for you, YouTube will get you to the 1965 song itself, performed two weeks earlier by the same musicians in 2016 at the Desert Trip concert, on October 14, the day after the announcement of the Nobel Prize. There you'll find the other Dylan singing a great, driving version, the guitars of Charlie Sexton and Donnie Herron unleashed as they follow Dylan's singing: "Once upon a time, you dressed so fine. . . ." Bob Dylan's world encompasses both songs and everything in between.

The band in these more recent concerts is now truly backing Dylan and his songs. Something has happened in the days since the performances of 2009–10, when Sexton returned to the band after leaving it in 2002, and Dylan seemed happy to showcase the new guitarist who was so right for the music that ended up on *Tempest*. Sexton is as good as ever, but he and all of the musicians are now there in the service of the band, the concert, the songs, and the singer. It is the songs, Dylan's performance, the integrity and completeness of the concert, and the story it tells, that are always the focus.

Dylan has scattered the songs of *Shadows in the Night, Fallen Angels,* and *Triplicate* across his set, mingling them with his own original songs, particularly those of the twenty-first century. He has given himself material for concerts for years to come, even if he makes no more records. In this setting the covers are no longer covers, no longer belong to anyone but Dylan, are part of his performative essence. Just as a line of Virgil in "Lonesome Day Blues" or Homer in "Early Roman Kings" no longer belongs to those ancient poets, but is stolen, a part of the song, these standards now belong to Dylan, precisely because they are heard in the arrangement and the performance of Dylan. And in the setting of a concert, individual songs become part of a larger, connected fabric. The new songs, as he told Love, "fall together to create a certain kind of drama." And now, integrated with the new, deliberately restricted setlists of these years, they participate in a larger drama, telling the story of Dylan's journey through the years.

Dylan at the end of the interview tries to explain to Love how he puts together a show:

> It starts like this. What kind of song do I need to play in my show? What don't I have? It always starts with what I don't have instead of doing more of the same. I need all kinds of songs—fast ones, slow ones, minor key, ballads, rumbas—and they all get juggled around during a live show. I've been trying for years to come up with songs that have the feeling of a Shakespearean drama, so I'm always starting with that.

"A certain kind of drama," and "the feeling of a Shakespearean drama." As before, Dylan goes back to Shakespeare, giving a foretaste of the Nobel acceptance speech he would deliver at the end of the following year, with which this book will close: "like Shakespeare, I too am often occupied with the pursuit of my creative endeavors and dealing with all aspects of life's mundane matters."

"I WAS THINKING IN TRIADS"

In 2010, Clinton Heylin introduced his chapter on the songs that would come out on *Modern Times* with a perceptive observation:

> In days of yore, Dylan had been something of a master when it came to producing trilogies of albums that served as the building blocks for a greater whole—witness the three acoustic albums he recorded between 1962 and 1964, the great electric trio of 1965–66, that (anti-)romantic trio he completed between November 1973 and July 1975, and the so-called religious trilogy released in the years 1979–81. The album he recorded in February 2006 proved to be the last volume of a trilogy of albums all hewn from the same pre-rock era of influences.

When *Tempest* came out in 2012, the trilogy seemed more easily to consist of that album and its two predecessors, *Love and Theft* (2001) and *Modern Times* (2006), with *Time Out of Mind* (1997) serving as the transitional comeback album. And Dylan continued with trilogies. In 2016, Dylan was working on the three-disc, thirty-song *Triplicate,* each of its sides more or less thirty-two minutes long. The songs could have fit on a double album, even though Dylan claimed for the thirty-two minutes "that's about the limit to the number of minutes on a long-playing record where the sound's most powerful." The real reason may have been somewhat different, more to do with formal expectations, with associations and connections to other triads in the old traditions in which he works. Flanagan also asked whether the title *Triplicate* brings to mind Frank Sinatra's trilogy of 1980, *Past Present Future*. "Yeah, in some ways, the idea of it," Dylan replied, adding, "I was thinking in triads anyway, like Aeschylus, *The Oresteia,* the three linked Greek plays. I envisioned something like that." I myself had wondered about a connection to Dante, whose trilogy, *The Divine Comedy,* has similar triadic perfection: *Inferno* 34 cantos long—Canto 1 is introductory, *Purgatory* and *Paradise* 33 cantos each, for a total of 100. That is also the tally of the satellite radio show *Theme Time Radio Hour* triad, with 100 episodes also across a triad of years from May 3, 2006, to April 15, 2009. But Dylan's mind indeed seems to have been more on drama, perhaps, as he says, on Aeschylus, but also on Shakespeare.

Why was Dylan thinking in triads in late 2016? I suggest it was because he was shaping his concerts of this period as dramatic trilogies, as he has more or less said, with the songs of the cover albums participating in the drama and movement of very specifically selected songs from his own arsenal. Starting in the fall of 2016, he limited his repertoire to the three great, or classic, periods of his musical career, all the songs, starting in Phoenix on October 16, right after the announcement of the Nobel Prize in Literature, coming from (a) 1963–66, (b) 1975, *Blood on the Tracks,* or (c) the post-1997 period, when the gift was given back with *Time Out of Mind*. These shows themselves also have a triadic essence, roughly three sections of six or seven songs. For the first section in every concert of the fall tour, the second and third songs were "Don't Think Twice, It's All Right" from 1963 and "Highway 61 Revisited" from 1965. This section has also featured a further triad in positions 5, 6, and 7: two songs from the recent "cover" albums, "Full Moon and Empty Arms" and "Melancholy Mood," in all twenty-six fall 2016 concerts framing and contrasting with the ominous, driving "Pay in Blood." That pattern continued in 2017, with new American standards stepping up to surround "Pay in Blood." "Desolation Row," also from 1965, was positioned later, the fifteenth song. At Clearwater, it featured Dylan sitting at the piano and turning to the audience as his facial expression seemed to act out the various masked characters in the song, itself a drama in its own right, whose faces, as the last verse puts it, Dylan had to rearrange as he gives "them all another name." The middle section of the triad was anchored by two songs from *Blood on the Tracks,* the middle classic period "Tangled Up in Blue" in tenth place and "Simple Twist of Fate" in the twelfth. "High Water Rising (For Charley Patton)" from 2001 and "Early Roman Kings" keep up the tempo and drive of this middle section, in which none of the American standards disrupted the focus on the run of six original Dylan songs, all with rich and complex poetic stories and visual imagery. In contrast, the third and final section as the evening draws to a close is characterized by a more melancholic mood, starting with the world-weary songs "Soon After Midnight" and "Long and Wasted Years," from the 2012 album *Tempest,* pinnacle of the long third classic period. Those two songs were joined by three from the American songbook, "I Could Have Told You," "All or Nothing at All," and "Autumn Leaves," totally at home and a fitting close to the main concert, with Dylan's voice clear and beautiful: "But I miss you most of all / My darling / When autumn leaves / Start to fall." This has nothing to do with Frank Sinatra, everything to do with the drama of the concert.

The performance at Clearwater was representative of all the fall shows, and the pattern continued in the spring and summer tours of 2017, though with "Tangled Up in Blue" the only representative of the seventies, as in some of the concerts from the previous fall. Here is the pattern, this specifically from Clearwater:

1. "Things Have Changed" (*The Essential Bob Dylan,* 2000)
2. **"Don't Think Twice, It's All Right"** (*The Freewheelin' Bob Dylan,* 1963)
3. **"Highway 61 Revisited"** (*Highway 61 Revisited,* 1965)
4. "Beyond Here Lies Nothin'" (*Together Through Life,* 2009)
5. "Full Moon and Empty Arms" (*Shadows in the Night,*2015)
6. "Pay in Blood" (*Tempest,* 2012)
7. "Melancholy Mood" (*Fallen Angels,* 2016)

8. "Duquesne Whistle" (*Tempest,* 2012)
9. "Love Sick" (*Time Out of Mind,* 1997)
10. **"Tangled Up in Blue"** (*Blood on the Tracks,* 1975)
11. "High Water (For Charley Patton)" (*"Love and Theft,"* 2001)
12. **"Simple Twist of Fate"** (*Blood on the Tracks,* 1975)
13. "Early Roman Kings" (*Tempest,* 2012)

14. "I Could Have Told You" (*Triplicate, Disc 1: 'Til The Sun Goes Down*, 2017)
15. "Desolation Row" (*Highway 61 Revisited,* 1965)
16. **"Soon After Midnight"** (*Tempest,* 2012)
17. "All or Nothing at All" (*Fallen Angels,* 2016)
18. **"Long and Wasted Years"** (*Tempest,* 2012)
19. "Autumn Leaves" (*Shadows in the Night,* 2015)

ENCORE

20. "Blowin' in the Wind" (*The Freewheelin' Bob Dylan,* 1962)
21. "Stay With Me" (*Shadows in the Night,* 2015)

The setlist for the shows in the fall of 2016 were highly distinctive, on paper, and as experienced in concert. What I heard in Clearwater on November 19 was completely new, completely different from what I had heard in the summer. Through ordering and repetition, what Dylan and his band were playing night after night were songs that added up to something that went beyond the sum of the parts, that had a certain narrative quality, told a connected story. That sum of the parts has a life of its own, in a sense is a life of Dylan, but also a life that his songs have constructed in the minds of those who have followed him and lived through his songs. He has, with the help of these songs and the support they lend to his own songbook, created a dramatic story of great beauty and power.

Then in June 2017 a funny thing happened, or rather two funny things. Dylan started the summer tour on June 13 with a three-night stand at the beautiful, renovated Capitol Theater in Port Chester, New York. After a diversion to the Firefly Music Festival in Dover, Delaware, the regular performances resumed in Wallingford, Connecticut, and points north in New England, New York, and Canada. With the—as always—kind help of the Bob Dylan office, to which I had sent a manuscript of this book on June 6, I had secured tickets for the second night in Port Chester on June 14, and in Providence, Rhode Island, a week later. I was looking forward to versions of the triadic setlist, but for my two shows and the three shows in between—but none before or after for the last year—the triad evaporated. "Tangled Up in Blue" was nowhere to be heard, taking with it any trace of the seventies, leaving only the first and last elements of the triad. "Tangled" came back at the next concert in Kingston, New York, and stayed for the remainder of the tour through the month of July, so restoring the triad.

As if by way of compensation, something else happened, starting with that same second Port Chester show. After the perennial opener "Things Have Changed," much to the delight of the crowd Dylan strapped on his guitar for a beautiful version of "To Ramona," one of the two mid-sixties songs that had anchored what was supposed in my mind to be first triad. "He didn't do that last night," said the man sitting next to me, who had earlier informed me "I'm Bob's lawyer." "That's right," I replied, "he's only picked it up once since October 13, the night of the Nobel announcement." That was pretty much when the triple structure of the concerts began, Dylan that night playing guitar for one song, "Simple Twist of Fate." I felt lucky to have experienced the sight and sound of the occasion eight months later in Port Chester. What I didn't know at the time was that we were witnessing a new performance triad, as Dylan again took up the guitar at the next two regular shows, for "Don't Think Twice, It's All Right," and then again for "It Ain't Me, Babe," so bringing back one of the many great songs on which the curtain had closed in those two nights in Rome in 2013. I don't know what all of this means. As Dylan would say, "you'll have to figure it out for yourself."

CURTAIN OPENING AND CLOSING

The Clearwater show began with "Things Have Changed," invariably the opener of the last four years, since April 5, 2013, in Buffalo, the first show of that year, the year in which performance setlists were radically restricted, with only twenty-six songs appearing five times or more in that year's eighty-five concerts, half the total number of songs played the previous year. The Oscar the song won for best original song in the movie *Wonder Boys*—or perhaps a facsimile of it—tours with Dylan and sits on top of the amplifier by his piano. Director Curtis Hanson said of the sound track, "Every song reflects the movie's themes of searching for past promise, future success and a sense of purpose." Much of the lyric quality of "Things Have Changed" is quite surreal: doing the jitterbug rag and dressing in drag; falling in love with the first woman he meets, putting her in a wheelbarrow and wheeling her down the street; having Mr. Jinx and Miss Lucy jumping in the lake. But much is quite clear and seems to capture what is happening in his music, including that surreal songwriting:

Lot of water under the bridge, lot of other stuff too
Don't get up gentlemen, I'm only passing through.

And then the close of the refrain, "I used to care, but things have changed," a phrase like so many in Dylan that stays relevant whatever the particular change his art is putting in play. In that sense the song is like "Ballad of a Thin Man," which came back as a closing song in the 2017 concerts, its refrain a challenge, once upon a time to folkies, now to new critics of the things that have changed, particularly the integration of songs from the American Songbook: "Because something is happening here / But you don't know what it is / Do you, Mister Jones?"

"Stay With Me" closed the Clearwater concert. I had heard that in Boston and was half-hoping for the wistful and beautiful "Why Try to Change Me Now?" my favorite from *Shadows in the Night*, the title recalling that of the opener. That song was written by Cy Coleman and Joseph McCarthy in 1952, and was covered by Sinatra in 1959 and Fiona Apple in 2009. This circle game defines, gives a frame to what comes in between. Things have always changed for Dylan, from the "Times" of his 1963 album, *The Times They Are A-Changin',* through what has happened at various phases of his life.

The things that have changed with his recent song list include creating a life in song, its dramatic qualities marking it as a self-contained performance, with a beginning, middle, and end, darkness onstage between the songs—or scenes—and no words other than the songs—just like a play. In the 2015 AARP interview, Dylan had Shakespeare on his mind, and at one point aligned plays and songs in ways that look ahead to the process he was working on with his concerts. Asked if he wished he had written some of the standards that were to become part of the drama of his concerts, he went off topic:

> I've seen *Othello* and *Hamlet* and *Merchant of Venice* over the years, and some versions are better than others. Way better. It's like hearing a bad version of a

> song. But then somewhere else somebody has a great version.

Perhaps not so off-topic. Songs, concerts, and plays finally come together.

DYLAN AND HIS FANS

I've got nothin' but affection for those who've sailed with me

—Bob Dylan, "Mississippi"

Bob Dylan may not say anything to his audience, but he is curious about who's out there; he's taking it all in. "What are you seeing from the stage?" Robert Love asked him:

> Definitely not a sea of conformity. People I cannot categorize easily. I see a guy dressed up in a suit and tie next to a guy in blue jeans. I see another guy in a sport coat next to another guy wearing a T-shirt. I see a woman sometimes in evening gowns, and I see punk-looking girls. I can see there's a difference in character, and it has nothing to do with age. I went to an Elton John show; there must have been at least three generations of people there. But they were all the same. Even the little kids. They looked just like their grandparents. It was strange.

There is something about being in a Dylan audience. The Clearwater crowd was on the older side, given the location, though there was a good mix of ages and there I even talked to a family with all three generations present, and it was true, they didn't look the same! The Boston crowd was pretty varied, quite a bit younger on average, baby boomers to millennials, and kids coming along for the ride.

The variety in Dylan's audience has more to do with the complexities of his long career, perhaps also his fame, though fewer seem go to a Bob Dylan concert to say they are going to a Bob Dylan concert. Until recently that could be a cause of serious distraction, with people reading their devices and texting, or talking during the songs. It was good to find a no-phones rule in Clearwater, also no coming back into the hall during a song. Again, that's what happens when you go to Shakespeare or the opera, another sign of what's going on with Dylan's concerts.

Particular phases and changes have brought in new followers, just as they have driven out others—and not just in the 1960s and '70s. It is true that if you liked country music, *Nashville Skyline* created a new appeal and attracted a different crowd from what went before, but Dylan wasn't touring in the years after that album came out, and I don't think people were really listening to that record as having much to do with country. The Christian period attracted some for the message, but that particular group would have been drifting away by *Shot of Love,* long gone by the time *Infidels* in 1983 seemed to renounce the solely Christian songs of 1979, explicit, unambiguous songs like "When He Returns," which Dylan abruptly stopped playing in November 1981. I have a friend who became a fan through the accident of playing *Under the Red Sky* to her small children when it came out in 1990. Why not? "Wiggle Wiggle," "Handy, Dandy," even "Under the Red Sky" work as well as anything for that purpose. Over my forty years of teaching I've encountered students who came on board at various stages, for the new music they discover on their own or the old music of their parents—or grandparents—who have long since stopped going to concerts but are still playing and listening to some songs.

The Blue Hills Bank Pavilion on the Boston waterfront is a pretty upscale place for a concert, a far cry from the minor-league baseball parks of a few years ago. As I do some of the time, I went on my own, and spoke to various people before the show. I ended up sitting next to a movie director from Los Angeles, who later sent me a bootleg version of "Highlands." In Clearwater there was a young woman singing and playing acoustic guitar in the outside bar area of Ruth Eckerd Hall. There I struck up a conversation with a couple named John and Sue, who had retired down there. "I'm one of the lucky ones who could," said John, who was originally from Danbury, Connecticut. Dylan had been there a year or two before. "How do you like the old standards?" I asked. Sue's response pretty much got it: "I don't mind what he sings. I'll listen to it whatever it is." In the show itself I was next to a young man from Croatia. It was his first show, but he knew the songs pretty well and was clearly enjoying it. He was with a German friend, who was more seasoned and had come over for Dylan and Neil Young at the Desert Trip the month before. If you follow the reviews on expectingrain.com, you see a lot of people planning vacations around Dylan. A year or two ago I met an Australian woman who was going to a number of shows while seeing something of the country, not a bad idea for a vacation. A number, like John and Sue in Clearwater, are locals. They just go when Dylan is in town, or close enough. I'll generally get to two or three a year myself.

Then there are some who just stay with the tour, all the way, day after day, show after show. I've met a few and have to say some of them remind me of the chess players I've seen in Harvard Square, playing chess all day long, decade after decade. You wonder how Dylan feels about those ones. For such a fan Dylan has become the Siren of the *Odyssey*: "no sailing home for him, no wife rising to meet him, no happy children beaming up at their father's face." And yet, there's a bit of that in many of us. We all flirt with it, just as Odysseus did. Ask my wife and children. And it goes in both directions. Dylan seems to need his fans as much as they need him—as he said, "it's the only place where I'm happy."

Part of the performance has to do with meeting people for the first time in a crowded bar or some burger place before the show, sharing stories, knowing you're going to add a chapter. For that evening, the rest of your life's activities, day jobs, worries about family, pretty much everything else, recedes and is replaced by a leveling, shared anticipation of how Bob and the band will be, what he will sing, even these days when you know, within a song or two, just what he'll be singing. The performance spills over into these moments, as precious as the shows, because they are part of the shows. This must be what it was like getting ready for the visit of the itinerant lyre players of ancient Greece, who would travel all around the Mediterranean playing to crowds, or talking with friends in ancient Athens in between the plays of a Greek tragic trilogy, say Aeschylus's *Oresteia,* where participating in a play by Aeschylus, Sophocles, or Euripides and shared Athenian citizenship were the same thing.

I felt a version of this after the Clearwater show. As I walked out into the warm November night, most people seemed to be smiling, glowing with warmth at what was an utterly perfect performance from start to finish. You know they'll be back. But for others that may have been the last waltz. I overheard three groups as I was leaving. One man, in his late thirties, far from sharing in the glow, was pretty angry: "No fucking country rock!" he exclaimed to the woman he was with. "He was like fucking Lawrence Welk! I slept in my car to do this. Prick!" Noticing my interest, she asked him to moderate his language. I wondered how *she* had liked the concert. Two women, also in their thirties or so, were more neutral: "Some of the words you could understand fine. If he wanted to, he could sing so you could understand everything." Her friend agreed. I thought Dylan's voice was magnificent, but if you didn't know a song like "High Water (For Charley Patton)" the lyrics might have been tough to pick up. Then there were two college kids, a little more knowledgeable it seemed, one of them complaining, "The only classic he's done in recent setlists is 'Blowin' in the Wind,' and he didn't even do that!" I couldn't resist intruding at this point, noting that he had in fact done the song, as the first of the two closers tonight as for some time. He didn't believe me, and I moved on, not wanting to press the point.

In all of these cases there was something at stake, something to do with memory, song, and shared human emotions and the joy, sorrow, or pain that is involved in listening to Bob Dylan. A long-dead grandfather; lost lover; wife, or husband, a friend who never made it through—a casualty perhaps of Vietnam, heroin, AIDS, Iraq, Afghanistan. Or to put it more simply, music and song are an essential part of being human, and particularly of being in the company of other humans. The music of our youth in particular stays with us. When it changes too much in performance, the singer may be doing things to those memories.

"He's just changed altogether," says the young English fan in Martin Scorsese's *No Direction Home*. "He's changed from what he was, he's not the same as what he was at first." "I didn't even recognize him," laments his friend. "Bob Dylan was a bastard in the second half," says another, referring to the electric backing of the Hawks from the tour of 1966, when Dylan did his famous half-acoustic, half-electric concerts. They should have known. He had already told them in March 1965, when *Bringing It All Back Home* came out, with the electric version of "Maggie's Farm":

> *Well, I try my best*
> *To be just like I am*
> *But everybody wants you*
> *To be just like them.*

Why has this been going on for more than fifty years? And why do people keep coming back? Because Dylan has become a classic, in fact always was, and that matters in the lives of the millions he has touched, even if he's moved on down the road from where they met him. That can disappoint those in search of where he was and they were on whatever occasion he mattered. And how does Dylan feel about that? "I used to care but things have changed," he sings. I say he cares, but that above all he cares, and always has cared, about his art and a vision that is the gift of genius. On to Stockholm!

CONCLUSION: SPEECHLESS IN STOCKHOLM

On October 1, 2008, the British paper the *Guardian* ran a headline above a photo of Philip Roth, NO NOBEL PRIZES FOR AMERICAN WRITERS: THEY'RE TOO PAROCHIAL. The source was Horace Engdahl, then permanent secretary of the Swedish Academy and one of the eighteen members whose job it is to select the winner of the Nobel Prize in Literature. "The U.S. is too isolated, too insular," said Engdahl. Himself fluent in six languages, he went on, "They don't translate enough and don't really participate in the great dialogue of literature." The response from writers in the United States had been as harsh as the statement, perhaps because Engdahl's words seemed to be an advance notice that there would be no American winner in 2008. David Remnick of the *New Yorker* told the Associated Press, "You would think that the permanent secretary of an academy that pretends to wisdom but has historically overlooked Proust, Joyce, and Nabokov, to name just a few non-Nobelists, would spare us the categorical lectures."

Sure enough, that year and for the next seven years, the drought stretching back to 1993, when Toni Morrison won the award, would continue, with a rich variety of non-American winners: French-Breton, Romanian-German, Spanish, Swedish, Chinese, Canadian, Ukrainian-Belarusian.

But by October 13, 2016, things had changed. Sara Danius, who had taken over as permanent secretary of the Nobel Committee, delivered the news to the applause and acclamation of the scores of journalists present for the occasion, in an elegant building that in an earlier age housed the Stockholm Stock Exchange. The text Danius read in Swedish, English, French, and German was as simple as it was meaningful and momentous: "The Nobel Prize in Literature for 2016 is awarded to Bob Dylan for having created new poetic expressions within the great American song tradition."

This moment was actually twenty years in the making. In 1996, two Dylan fans in Norway, journalist Reidar Indrebø and attorney Gunnar Lunde, contacted the office of Beat poet Allen Ginsberg, hoping for support for the nomination of Bob Dylan. Nobel nominations are only accepted from members of the Swedish Academy itself or of other similar academies, from professors of literature and language, from past Nobel laureates, or from presidents of literary societies around the world. Ginsberg met none of these criteria, but he had decided to anoint Dylan as his successor, one poet thus recognizing the artistic genius of another who would replace and eclipse him. Ginsberg's office put the wheels in motion by contacting a professor at the Virginia Military Institute named Gordon Ball, who was the author of three books on Ginsberg, and who had been a fan of Bob Dylan since first seeing him perform at the historic Newport Folk Festival in 1965. As a result of this chain of communication, beginning in 1996, and then again every year afterward, Ball had been nominating Bob Dylan for the Nobel. Two decades later, in 2016, the message had finally gotten through. As one commenter put it, "Looks like the Nobel Committee has gone electric."

The reaction to the announcement from the literary community was swift and uncompromising, and a mix of celebration and detraction. Stephen King, Joyce Carol Oates, and Salman Rushdie immediately hailed the choice. Rushdie, who may well have been a candidate himself, was unstinting, quoted in the *New York Times* as saying that "from Orpheus to Faiz, song and poetry have been closely linked," and calling Dylan "the brilliant inheritor of the bardic tradition," with a further punctuation, "Great choice." King declared himself "ecstatic" and called the choice "a great and good thing in a season of sleaze and sadness"—this in the difficult and tawdry weeks leading up to the 2016 U.S. presidential election. Andrew Motion, poet laureate of the UK, 1999-2000, considered the award "a wonderful acknowledgement of Dylan's genius: for 50 and some years he has bent, coaxed, teased and persuaded words into lyric and narrative shapes that are at once extraordinary and inevitable."

Those who objected were in the minority, their objections often definitional: Dylan might be a good singer-songwriter, but without music his words could not stand on their own, and thus were not poetry or literature. So, for instance, Irish literary critic Edna Longley called the award "a ridiculous decision, and an insult to real poets." But by the time of the ceremony in Stockholm, naysayers in the media and blogosphere had largely been silenced. In the words that Engdahl read at the ceremony, the awarding of the Nobel Prize "was a decision that seemed daring only beforehand and already seems obvious."

The Nobel Prize ceremony, also triadic, a play in three parts, took place on December 10, 2016, and the formal address that evening was delivered by the same Horace Engdahl who had called American writers parochial eight years earlier. The speech sounded very much like the work of a committee, with various threads that somehow all came together, in many ways reflecting the complexity of the phenomenon that is Bob Dylan. The address began with the opening question: "What brings about the great shifts in the world of literature?" The answer got to the heart of the matter: "Often it is when someone seizes upon a simple, overlooked form, discounted as art in the higher sense, and makes it mutate." The committee had also closely engaged with the question of whether song can be literature. Maybe they even debated or disagreed over the issues, which has long seemed irrelevant to many, as the address went on to note:

> In itself, it ought not to be a sensation that a singer/songwriter now stands recipient of the literary Nobel Prize. In a distant past, all poetry was sung or tunefully recited, poets were rhapsodes, bards, troubadours; "lyrics" comes from "lyre."

Throughout Engdahl's address, we are inevitably hearing the voices of various members of the academy, as well as of Gordon Ball from his annual nominations, and of Dylanologists whose writings had been brought to the attention of those voting.

Engdahl went on to talk of the many qualities that had persuaded the committee: the creativity that begins with imitation, the dazzling rhymes "scarcely containable by the human brain," Dylan's love songs, his eclipse of those in whose tradition he sang—Woody Guthrie, Hank Williams, Blake, Rimbaud, Whitman, Shakespeare, his bringing back of a poetic language "lost since the Romantics." For the Nobel Committee, Bob Dylan mattered because of his creation of an art that compelled them to see it as literature of the highest order:

> By means of his oeuvre, Bob Dylan has changed our idea of what poetry can be and how it can work. He is a singer worthy of a place beside the Greeks' *aoidoi* ["poet-singers"], beside Ovid, beside the Romantic

visionaries, beside the kings and queens of the Blues,
beside the forgotten masters of brilliant standards.

PATTI SMITH COVERS "A HARD RAIN'S A-GONNA FALL"

Next came the second act, starring Patti Smith, a sort of channel to Dylan, who had opted not to attend the award ceremony. Bob Dylan's absence in Stockholm, and the fact that it took him weeks to respond when the prize was first announced, are matters of speculation. My guess is that he just couldn't see himself in that august room; it wasn't his thing. So Patti Smith singing Dylan's "A Hard Rain's A-Gonna Fall" was as close as we were going to get that night. Anyone who saw that performance was a live witness to why Dylan matters. The modern significance of his work was seemingly encapsulated by this one song, written more than half a century earlier in Greenwich Village by a twenty-one-year-old. The song's line "Heard the song of a poet who died in the gutter" may well have come from the reality of Dylan's life in the Village in those days, as he sang in cafés for hamburgers and spare change and slept on couches. Dylan wrote "A Hard Rain" in the summer of 1962 and first performed it on September 22, in the weeks before the Cuban missile crisis of October 16–28, the closest the world came to all-out nuclear war. When the song was released on May 27, 1963, on Dylan's first original album, *The Freewheelin' Bob Dylan,* it was naturally assumed that the "hard rain" was the rain of nuclear bombs that had threatened a few months before. Indeed, by then there was no reason *not* to connect it to such events—though ultimately the language of the song, and the absence of defining and limiting topical, geographical, or chronological elements, make it a song for any time. Dylan has said of this and by extension of all song, "it doesn't really matter where a song comes from. It just matters where it takes you."

Each verse of Dylan's song begins with a variant of a line from a seventeenth-century Anglo-Scottish ballad called "Lord Randall." In the original ballad, the singer addresses a character named Lord Randall: "Oh where ha you been, Lord Randall, my son, / And where ha you been, my handsome young man." But in Dylan's song he substitutes "my blued-eyed son" for "Lord Randall," thus allowing the song to be addressed to the strikingly blue-eyed Bob Dylan himself.

If we interpret the lyrics as a sort of call-and-response between Dylan the singer and Dylan the addressee, the song transforms from a narrative ballad to a cry of warning that Dylan has to offer to a world gone wrong:

And I'll tell it and think it and speak it and breathe it
And reflect it from the mountain so all souls can see it

Each of the scenes the blue-eyed boy encounters in the final verse is as vivid now as it was back in the 1960s:

Where the people are many and their hands are all empty
Where the pellets of poison are flooding their waters
Where the home in the valley meets the damp dirty prison
Where the executioner's face is always well hidden
Where hunger is ugly, where souls are forgotten
Where black is the color, where none is the number

In a piece she wrote for the *New Yorker* in December 2016, Patti Smith recalls how she came to sing "A Hard Rain" at the ceremony. She had first heard it in 1963, the year the album came out, when she was sixteen. According to Smith, her mother, a waitress, had bought the album for her secondhand, using her tip money. Smith described it as "a song I have loved since I was a teenager, and favorite of my late husband"—guitarist Fred "Sonic" Smith.

"I saw a black branch with blood that kept drippin'," Smith sang, halfway through the second verse. At that moment, the camera focused on presenter of the Nobel Prize in Physics, Thors Hans Hansson, a distinguished gray-haired, bearded gentleman. His formal white tie and tails might have led one to think he would disapprove of a Bob Dylan song being performed at the Nobel ceremony, following the Royal Stockholm Philharmonic Orchestra's rendition of Jean Sibelius's *Serenade,* from the *King Christian II Suite*. But then Hansson's lips started moving as he sang to himself, "blood that kept drippin'." Clearly the theoretical physicist was at that moment just another Dylan fan. And if one expected to see doubt or disapproval on the faces of others in attendance, either at the novelty of the award or at the failure of Dylan to come to their ceremony, the truth was far from it, as Dylanologist David Gaines reported:

> The Swedish Minister of Culture, a striking woman in a red dress, cried throughout the song. When Smith closed, the royals and the other 1,250 people looked toward her and applauded. My Swedish friends with whom I watched the broadcast (it is the most widely watched program in Sweden every year) told me, "We have never seen such applause before."

At this point the camera moved back toward the stage as Smith stumbled over the lyrics, halfway through the second verse of a song that she had been singing with such power. She struggled to find her place, eventually giving up and turning to the orchestra: "Sorry. I'm sorry. Could we start that section again?" She looked out at the audience: "I apologize. Sorry. I'm so nervous." In response, the hall seemed to echo with applause, restoring her confidence. She picked up where she had left off and finished the song beautifully, from the crowd's perspective. As Smith wrote of the experience in the *New Yorker:*

> From the corner of my eye, I could see the huge boom stand of the television camera, and all the dignitaries

> upon the stage and the people beyond. Unaccustomed to such an overwhelming case of nerves, I was unable to continue. I hadn't forgotten the words that were now a part of me. I was simply unable to draw them out.

"Patti Smith botches Nobel tribute to absent Bob Dylan," proclaimed the *New York Post* on December 11, a putdown of both artists. The *Post* missed the point. In that lapse you see the frailty and humanity of the singer, as she goes on to complete the song, but you also see her resilience. Through Smith's performance, we witnessed the lyrics of "A Hard Rain" come to life, showing us what it means to fall, as we've all fallen, and to get up and struggle on. Smith was performing for an audience that included the king of Sweden and other royals, as well as scientists and academics and members of the Swedish Academy and high society. The roomful of men in white tie and tails and women in evening gowns lent a strict formality to the occasion, with differences and individuality concealed behind evening dress. Those six words—"I apologize. Sorry. I'm so nervous"—were so honest and vulnerable that they shifted the tone and canceled out any differences in the room, between physics and folk song, chemistry and rock, medicine and popular culture.

In the early sixties, years that had students hiding under their desks in nuclear war drills, the song's lyrics had taken listeners to the threat of Cold War missile attacks. On the night of Smith's performance, fifty-five years after the song was written, its lyrics conjure up new associations: people "whose hands are all empty" remind us of an inequality throughout the world that seems endless; "pellets of poison" might evoke the poison of electoral politics in 2016, or the environmental consequences; "damp dirty prison" might suggest mass incarceration in the United States or Kalief Browder, the black teenager held for three years without trial and in solitary confinement at New York's Rikers Island, dead by his own hand at twenty-two. As for the "always well-hidden" executioner's face, it's not now the hooded executioner, but maybe a uniformed figure in a dimly lit room, deploying a drone far from its target—like playing a video game. What other song from 1962 still works the way "Hard Rain" does?

It is a song of indignation, but also a song of resolve—"I'll tell it and think it and speak it and breathe it." Patti Smith showed that as she recovered and finished the song, which mattered in that moment, as surely it did back when Dylan wrote it. It is also a song of beauty. The tears that many shed as they watched Smith came from a place of human desire, or need, for what is beautiful.

DYLAN'S NOBEL BANQUET SPEECH

It had initially taken Dylan more than two weeks to respond to the announcement of the award. Sara Danius had seemed patient four days after the initial announcement, as she told the

Telegraph on October 17. "Right now we are doing nothing," said Danius. "I have called and sent emails to his closest collaborator and received very friendly replies. For now that is certainly enough." In the two weeks since the October 13 announcement, Dylan performed eleven concerts in eight states, from California to Mississippi. He may just have been focused on what matters most to him: performing his songs. Eventually, on October 28, the Swedish Academy put out a press release under the heading "The Call from Bob Dylan":

> "If I accept the Prize? Of course." . . . This week Bob Dylan called the Swedish Academy. "The news about the Nobel Prize left me speechless," he told Sara Danius, Permanent Secretary of the Swedish Academy. "I appreciate the honor so much."

The next day the journalist Edna Gunderson published an interview with Dylan in the *Telegraph,* scheduled in connection with an upcoming exhibition of Dylan's artwork in London. Gunderson, quoting Danius, asked Dylan how he felt about the permanent secretary's connecting his songs to poetic texts of classical antiquity:

> "If you look back, far back, 2,500 years or so," she [Danius] has said, "you discover Homer and Sappho, and they wrote poetic texts that were meant to be

> listened to, they were meant to be performed, often together with instruments, and it's the same way with Bob Dylan. But we still read Homer and Sappho . . . and we enjoy it, and same thing with Bob Dylan. He can be read, and should be read."

Dylan responds with some hesitation:

> "I suppose so, in some way. Some [of my own] songs—'Blind Willie,' 'The Ballad of Hollis Brown,' 'Joey,' 'A Hard Rain,' 'Hurricane,' and some others—definitely are Homeric in value." Of this Gunderson says, "He has, of course, never been one to explain his lyrics. 'I'll let other people decide what they are,' he tells me. 'The academics, they ought to know. I'm not really qualified. I don't have any opinion.'"

With this answer from Dylan, who is almost invariably silent on the meaning of his songs, the door opened a little, as he acknowledged that some of them might legitimately be considered "Homeric." The fact that he chose to compare "Blind Willie [McTell]," a song about a blind blues singer, to Homer, another blind poet, was typical of Dylan's sense of humor. In reality, he had, night after night since news of the award was released, been singing the words of Homer's Odysseus from "Pay in Blood" and "Early Roman Kings"—truly Homeric songs—

but he wasn't about to let that out. Dylan goes back to Homer, but he also differs gently but surely from Danius in his Nobel lecture, honest to the end: "songs are unlike literature. They're meant to be sung, not read." So much for that question.

Eventually, on November 17, the Associated Press reported that Dylan had told the Swedish Academy that "he wishes he could receive the prize personally, but other commitments make it unfortunately impossible." He did, however, send in an acceptance speech, a condition of receiving the Nobel. Read by Azita Raji, the U.S. ambassador to Sweden, it was a tour de force, a demonstration to anyone who needed it that they had the right man, coming across as sincerely grateful, and marked by elegance, wit, and humor. Raji delivered Dylan's prose to the banqueters:

> Good evening, everyone. I extend my warmest greetings to the members of the Swedish Academy and to all of the other distinguished guests in attendance tonight.

The same gracious tone continues:

> I'm sorry I can't be with you in person, but please know that I am most definitely with you in spirit and honored to be receiving such a prestigious prize.

Then things get interesting as Dylan's speech begins to reveal its artistic purposes:

> From an early age, I've been familiar with and reading and absorbing the works of those who were deemed worthy of such a distinction: Kipling, Shaw, Thomas Mann, Pearl Buck, Albert Camus, Hemingway. These giants of literature whose works are taught in the schoolroom, housed in libraries around the world and spoken of in reverent tones have always made a deep impression. That I now join the names on such a list is truly beyond words.

A few weeks earlier, Dylan had said that the award had left him "speechless," and now the sentiment was "beyond words"—but this at the end of some fairly specific words from the man of whom Joan Baez sang, "you who're so good with words, and at keeping things vague." He chooses to cite six specific writers, now fellow laureates, as making an impression on him, George Bernard Shaw but no W. B. Yeats or Seamus Heaney. Hemingway but no William Faulkner or even John Steinbeck.

Nor does Dylan mention T. S. Eliot, who famously appeared along with Ezra Pound in his 1965 song "Desolation Row," whose words borrowed, or stole, from Eliot's first poem *The Love Song of J. Alfred Prufrock*, as noted at the beginning of this book. In his memoir, *Chronicles: Volume One,* Dylan makes a distinction: "I never did read him [Pound]. I liked T. S. Eliot. He was worth reading." It is curious that Dylan failed in his acceptance speech to mention the poet whose work he must have highly regarded, and who transformed poetic tradition in the 1920s in a way that's comparable to Dylan's transformation of songwriting traditions forty years later. I suspect that Dylan left Eliot out through real modesty. He knew what Eliot had done for English literature and was not quite ready to put himself on that pedestal. Even if that is where he belongs.

Dylan's tone is modest, but his typical allusiveness and humor are never far off:

> If someone had ever told me that I had the slightest chance of winning the Nobel Prize, I would have to think that I'd have about the same odds as standing on the moon.

Is Dylan here alluding to Seamus Heaney's 1995 Nobel acceptance speech, which also compared the prize to venturing into space? Or was "standing on the moon" just a nod to Robert Hunter, who wrote the Grateful Dead song of that title with Jerry Garcia, and who also cowrote most of the songs of *Together Though Life* and the first song on *Tempest,* "Duquesne Whistle," with Dylan?

Dylan's next sentence also shows that humor is in the air:

> In fact, during the year I was born and for a few years after, there wasn't anyone in the world who was considered good enough to win this Nobel Prize. So, I recognize that I am in very rare company, to say the least.

No one was good enough in the literature category in 1941, the year Dylan was born? Or could it be that Sweden and the Swedish Academy had more pressing things going on in the difficult war years from 1940 to 1942, and that's why they awarded no prizes in any category during this period? Dylan, a historian at heart, knew this full well. Good one, Bob!

What about Dylan and Shakespeare? People have been connecting the two names for years, and rightly so. With modesty, style, and wit, Dylan gave us an answer in his address:

> I was out on the road when I received this surprising news, and it took me more than a few minutes to properly process it. I began to think about William Shakespeare, the great literary figure. I would reckon he thought of himself as a dramatist. The thought that he was writing literature couldn't have entered his head. His words were written for the stage. Meant

> to be spoken not read. When he was writing Hamlet, I'm sure he was thinking about a lot of different things: "Who're the right actors for these roles?" "How should this be staged?" "Do I really want to set this in Denmark?" His creative vision and ambitions were no doubt at the forefront of his mind, but there were also more mundane matters to consider and deal with. "Is the financing in place?" "Are there enough good seats for my patrons?" "Where am I going to get a human skull?" I would bet that the farthest thing from Shakespeare's mind was the question "Is this literature?"

The point Dylan makes is a serious one, and it's one that draws an interesting parallel between Shakespeare and himself as both concerned with performance, and not "literature." But the fact that he makes it with humor, with poor Yorick's skull carrying the punch line, removes any possibility for self-importance. "Do I really want to set this in Denmark?" . . . "Where am I going to get a human skull?" Dylan loves the three-part parallel question, of which there are two sets here, and he will return to this device later in his speech. It belongs to the world of rhetoric and persuasive speech, going back to the Greeks and Romans, and it is at the core of "Blowin' in the Wind," the song he positioned in first place on his first original album: "How many roads . . ." But it also belongs to the world of jest, vaudeville, and the punch line, worlds never far from Dylan's mind. Before coming back to the triple question, Dylan first turns to his own song in performance, to what he has always cared most about:

> Well, I've been doing what I set out to do for a long time, now. I've made dozens of records and played thousands of concerts all around the world. But it's my songs that are at the vital center of almost everything I do. They seemed to have found a place in the lives of many people throughout many different cultures and I'm grateful for that.

Dylan's expression of gratitude for the place he and his songs have found in the hearts of millions gives a rare glimpse, seldom seen in performance, of his appreciation for his fans. As he put it on the 1997 song "Mississippi," "I've got nothing but affection for all those who've sailed with me," or in the words of the cover song with which he has closed many concerts in recent years, "Stay with me."

Soon his speech elegantly loops back to Shakespeare, and their shared concern with performance over creating high literature. This time, as Dylan returns to the three-part question, there is no joke. These are the questions he has been concerned with for years, the questions that go to the heart of his art:

> But, like Shakespeare, I too am often occupied with the pursuit of my creative endeavors and dealing with all aspects of life's mundane matters. "Who are the best musicians for these songs?" "Am I recording in the right studio?" "Is this song in the right key?" Some things never change, even in 400 years.
>
> Not once have I ever had the time to ask myself, "Are my songs literature?"

In closing, Dylan's speech comes full circle as he accepts the academy's judgment and honors their answer, not his, to this question, which the world had been debating for the last two months:

> So, I do thank the Swedish Academy, both for taking the time to consider that very question, and, ultimately, for providing such a wonderful answer.
>
> *My best wishes to you all,*
>
> **BOB DYLAN**

The verdict is in. Yes, Dylan's work is literature, in an expansive rather than a limiting sense of the word. He is rightly the holder of the Nobel Prize in Literature for 2016, and you can add his speech to the rest of his masterpieces. But at the same time, Dylan's award gives us reason to call into question the way we define "great literature" in modern society. If literature is only something that gets written down, valued by literate societies, preserved in libraries, read in solitude and taught in schools, much of the material that went into Dylan's songs—at least until his songwriting in the twenty-first century—is not itself great literature. And yet, how do we classify artistic work that springs from the oral tradition of the blues and folk, and that survives and thrives in illiterate and semiliterate cultures because what those traditions preserve is something human communities need? Is this work not literary until a Dylan, or a classical poet like Homer, comes along and the song is written down? It is a mark of Dylan's art, and of his genius, that the song he has created—and performed—is something that matters as much as the more conventionally literary traditions that, like his song, convey solace, joy, and sadness to humanity.

DYLAN'S NOBEL LECTURE

On June 5, 2017, days before the deadline of June 10, the Swedish Academy announced that Bob Dylan had delivered a video of the 2016 Nobel Lecture in Literature, taped in Los Angeles on June 4. The week before, as reported in the *New York Times* on June 7, jazz pianist Alan Pasqua, who had briefly played for Dylan in the late seventies, was contacted by Dylan's business manager:

> "Have you ever watched those old clips of Steve Allen interviewing people, when he plays the piano?" And I was like, yeah! And he said, "Well, we need some of that kind of music. You know, not really melodic, not cocktail, not super jazzy, but sort of background-y piano music."

Pasqua obliged, and the music can be heard gently and unobtrusively in the background. Its presence should be one clue that the lecture would not be conventional. It is however highly informative, with Dylan talking about Buddy Holly at the "dawning of it all," the singer whose intertwining of country western, rock and roll, and rhythm and blues into one genre showed the seventeen-year-old Bob Zimmerman what was possible, as did the "beautiful melodies and imaginative verses" Holly created. In the drama of Dylan's lecture—clearly a work of creative imagination as well as actual truth—right around the time Holly died in the plane crash of 1959, someone hands him a Leadbelly record. This transports him "into a world I'd never known," the world of work songs, prison songs, gospel, the blues. A leaflet that came with the Leadbelly record introduced him to Sonny Terry, Brownie McGhee, the New Lost City Ramblers, and Jean Ritchie, all on the same label as Leadbelly, and therefore bound to be worth listening to, he informs us. A neat way of describing or constructing his genesis as folksinger!

Dylan then moves on to a fascinating description of how

he gained a mastery over the "vernacular" of the early folk artists by singing the songs: "You internalize it. You sing it in the ragtime blues, work songs, Georgia sea chanties, Appalachian ballads and cowboy songs. You hear all the finer points, and you learn the details." These are important observations for those wishing to understand how Dylan's songwriting genius came to be, the experience and observation that go along with his imagination. There follows a magnificent paragraph in which the lecture enters into the worlds of these old traditions, and the use of the pronoun "you," pointing to Bob Dylan the songwriter, anticipates what we already saw with respect to the *Odyssey* at the end of the lecture, where "you" included Dylan and Odysseus:

> You know what it's all about. Takin' the pistol out and puttin' it back in your pocket. Whippin' your way through traffic, talkin' in the dark. You know that Stagger Lee was a bad man and that Frankie was a good girl. You know that Washington is a bourgeois town and you've heard the deep-pitched voice of John the Revelator and you saw the *Titanic* sink in a boggy creek. And you're pals with the wild Irish rover and the wild colonial boy. You heard the muffled drums and the fifes that played lowly. You've seen the lusty Lord Donald stick a knife in his wife, and a lot of your comrades have been wrapped in white linen.

Just as he becomes Odysseus later in the lecture—"You too have had drugs dropped in your wine"—so too here he has entered into the folk songs and ballads which he has hardwired and whose world he inhabits. This is what it means to live inside the world of literature and song. Dylan's words are crowded with lines of songs: "Stagger Lee was a bad man" ("Stagger Lee"), "Frankie was a good girl" ("Frankie and Albert"), "the fife that played lowly" ("Streets of Laredo"), and the absurdism of what Dylan's songwriting can do is on full display: "you saw the *Titanic* sink in a boggy creek." He may even be having a little political fun. In the song "Mattie Groves," which goes back to the early seventeenth century and is known by various other titles, Mattie, aka "Little Musgrave," goes to bed with Lady Barnard. Lord Barnard catches the two *in flagrante delicto*, kills Mattie, and then kills his wife after she expresses a preference for Mattie—sounds like the situation in the 2012 song "Tin Angel." Mattie would rightly be called the "lusty" one in this ballad. Lord Barnard, who has a number of variant names, including "Lord Donald" is the cuckold. But Dylan's "lusty Lord Donald" may have appealed to him for contemporary political reasons.

The bulk of the lecture focuses on three books out of many, read back in "grammar school," from which Dylan had acquired "principles and sensibilities and an informed view of the world." Along with the *Odyssey*, he names Herman Melville's *Moby-Dick* (1851) and Erich Maria Remarque's *All Quiet on the*

Western Front (1929). For the former Dylan particularly focuses on character and on the book's "scenes of high drama and dramatic dialogue," with a rousing conclusion at the end of the hunt for the great white whale. His thoughts here seem to revolve around faith and creeds, resurrection and survival, and in the most eclectic way: "Everything is mixed in. All the myths: the Judeo-Christian Bible, Hindu myths, British legends, Saint George, Perseus, Hercules—they're all whalers." He also explores how to respond to the hand fate deals us, how "different men react in different ways to the same experience." The lesson for Dylan is in the contrast between Ahab, obsessed and driven by Moby, and Captain Boomer, who "lost an arm to Moby. But he tolerates that, and he's happy to have survived. He can't accept Ahab's lust for vengeance." The narration here is straightforward, without that autobiographical use of "you," but also deals in masks and personas:

> We see only the surface of things. We can interpret what lies below any way we see fit. Crewmen walk around on deck listening for mermaids, and sharks and vultures follow the ship. Reading skulls and faces like you read a book. Here's a face. I'll put it in front of you. Read it if you can.

He ends as Melville began, with Ishmael, the novel's narrator, who like Dylan the songwriter, is the one who makes the masks: "Ishmael survives. He's in the sea floating on a coffin. And that's about it. That's the whole story. That theme and all that it implies would work its way into more than a few of my songs."

While *Moby-Dick* for Dylan offers ways of surviving, along with choices to make, Dylan's response to another book that worked its way into his songs, *All Quiet on the Western Front*, a "horror story," is very different. At first the use of the second person "you" seems to refer to the audience, also to Dylan himself, as he begins:

> This is a book where you lose your childhood, your faith in a meaningful world, and your concern for individuals. You're stuck in a nightmare. Sucked up into a mysterious whirlpool of death and pain.

The German novel, written in the aftermath of the first world war, and so effective as an antiwar novel that the Nazis burned it and banned it from a Germany that was moving toward the catastrophe of the next world war, is until its last page narrated entirely in the first person. The narrator and lead character is Paul Bäumer, through whose eyes we see the human degradation brought about by life in the mud and rat- and corpse-infested trenches of the "war to end all wars."

Dylan's narration, almost a third of the entire lecture, is astonishing. From the very next sentence, it becomes clear that the identity of the "you" is in fact not Dylan or us, but Paul himself:

> You're defending yourself from elimination. You're being wiped off the face of the map. Once upon a time you were an innocent youth with big dreams about being a concert pianist. Once you loved life and the world, and now you're shooting it to pieces.

For more than six minutes, piano softly playing in the background, the Nobel lecture becomes a talking blues, a distillation of the entire book, turning Remarque's first person into a second person, as Dylan addresses the young soldier-narrator. The detail is relentless, as is the book that it distills, with Dylan's poetic powers in full view:

> More machine guns rattle, more parts of bodies hanging from wires, more pieces of arms and legs and skulls where butterflies perch on teeth, more hideous wounds, pus coming out of every pore, lung wounds, wounds too big for the body, gas-blowing cadavers, and dead bodies making retching noises. Death is everywhere. Nothing else is possible. Someone will kill you and use your dead body for target practice. Boots, too. They're your prized possession. But soon they'll be on somebody else's feet.

Through the inhumanity depicted by the book in Dylan's empathetic and brilliant retelling there shines a ray of humanity, especially right before shrapnel hits Paul in the head and kills him. In the book the narration at that moment shifts from first person to third: "He fell in October 1918, on a day that was so quiet and still on the whole front that the army report confined itself to the single sentence: All quiet on the Western front," then proceeding dispassionately to relate the death of the person whose voice has led us throughout the book: "he had fallen forward and lay on the earth as though sleeping . . ." Dylan ends by switching from second to first person, with an epitaph to Paul followed his own reflection:

> You're so alone. Then a piece of shrapnel hits the side of your head and you're dead. You've been ruled out, crossed out. You've been exterminated. I put this book down and closed it up. I never wanted to read another war novel again, and I never did.

Like the book, Dylan's lecture is also a searing indictment of the old who make the wars and send the young to their deaths in battle: "You've come to despise that older generation that sent you into this madness, into this torture chamber." For Dylan, as for many of his generation, these thoughts go back to Dwight Eisenhower's 1961 warnings about the proliferation of the "military-industrial complex," issued the very week the twenty-year-old Dylan arrived in New York City in that frigid January. He may not have read another war novel again, but he didn't need to. He had all that he needed to write "Masters of War," the greatest anti-war song ever written. Three or four years after reading *All Quiet* at Hibbing High, Dylan finally found in Jean Ritchie's song "Nottamun Town" the melody for his version of the novel:

> *You hide in your mansions*
> *As young people's blood*
> *Flows out of their bodies*
> *And is buried in the mud.*

"SEAL UP THE BOOK"

It is time to "seal up the book and not write any more," as Dylan sings in "Tryin' to Get to Heaven," from *Time Out of Mind*. But this is a hazardous business, since Dylan's work continues, and words can come back to haunt you. For all I know, just as this book goes to press, Dylan's camp may announce that a sequel to *Tempest* is in the works, or is even about to be released, my triad may have disappeared, the songs left in Rome in November 2013 may come crowding back into the setlist.

Two of the foremost Dylan scholars out there, Michael Gray and Clinton Heylin, were quite aware of the predicament. Gray's *Song and Dance Man III,* published in 2000, ended with a lamentation of the state of Dylan's art at the time, with Gray fervently hoping that he "refuses to settle for this comfortable descent, in an apparently inevitable smooth arc, into being a performer and writer of less and less artistic power." But Gray also concluded by allowing that there might be an ascent, albeit with faint praise ("faltering step") for the extended miracle that started happening in 1997:

> Impending old age is itself good raw material. Instead of clinging to his back catalogue, he could voice his real concerns, as once he did, and be glad to have an audience: 1997's "Time Out of Mind" is a faltering step along this path.

Only seven of twenty-one songs from the 2017 tour existed in Dylan's songbook when Gray wrote that. The path from 2000 has been long and wide.

The approach of Clinton Heylin, to whom scholars and fans of Dylan, myself not least, are most in debt, was to continue supplementing his 1991 *Behind the Shades* biography, its 498 pages growing to 780 pages in the 2001 "Revisited" volume, and 902 in the 20th Anniversary Edition in 2011, getting us down to 2010, so not covering *Tempest*. This is not to count his five other books on Dylan, especially the two-part song-by-song histories, *Revolution in the Air* (2009) and *Still on the Road* (2010). By the time you are reading this book, a sixth, his treatment of the gospel years, *Trouble in Mind,* will have been added.

So I prefer to give Dylan the last word. In his most recent interview to date, with writer and TV executive Bill Flanagan on March 22, 2017, Dylan to my mind confirms many of the aspects of his art that I have engaged in this book. He doesn't use the word intertextuality—why should he?—but he is talking about the phenomenon when he says, "Try to create something original, you're in for a surprise." Instead, like the poets you have met—maybe for the first time—in this book, he has other ways of creating. In the interview, Flanagan asks him a question, which sounds a lot like a plant by Dylan: "People yell about plagiarism . . . but it has always gone on in every form of music, hasn't it?" Dylan replies:

> I'm sure it has, there's always some precedent—most everything is a knockoff of something else. You could have some monstrous vision, or a perplexing idea that you can't quite get down, can't handle the theme. But then you'll see a newspaper clipping or a billboard sign, or a paragraph from an old Dickens novel, or you'll hear some line from another song, or something you might overhear somebody say just might be something in your mind that you didn't know you remembered. That will give you the point of approach and specific details. It's like you're sleepwalking, not searching or seeking; things are transmitted to you. It's as if you were looking at something far off and now

> you're standing in the middle of it. Once you get the idea, everything you see, read, taste or smell becomes an allusion to it. It's the art of transforming things. You don't really serve art, art serves you and it's only an expression of life anyway; it's not real life.

Flanagan also asks Dylan about the world that has been lost since the time of the writers of the Great American Songbook. His response refers to those songs but also to his art in general and to the ways that art has worked for him for a long time:

> You can still find what you're looking for if you follow the trail back. It could be right there where you left it—anything is possible. Trouble is, you can't bring it back with you. You have to stay right there with it. I think that is what nostalgia is all about.

I've been on that trail for a long time, engaging not just with the world of the songs that Dylan wrote in the mid-twentieth century, but going way back, all the way to Homer, Virgil, and Ovid. And Bob Dylan, the supreme artist of the English language of my time, has been on that same trail, going back to ancient times to mine material for his work, and making it about the here and now. In 1997, the last song of the comeback album *Time Out of Mind* treats that nostalgia, being "there in my mind" in the Highlands of Robert Burns, yes, but as the next twenty years would show, in even more ancient worlds:

> *Well my heart's in the Highlands at the break of day*
> *Over the hills and far away*
> *There's a way to get there and I'll figure it out somehow*
> *But I'm already there in my mind*
> *And that's good enough for now.*
>
> —Bob Dylan, "Highlands"

Good enough for me, too.

ACKNOWLEDGMENTS

I have many people to thank, beginning with Jeff Rosen and his colleagues in Bob Dylan's office, who have been so helpful and generous to scholars of Dylan over the years, myself included. Along with Jeff Rosen, I thank David Beal, Larry Jenkins, Debbie Sweeney, Raymond Foye, and Lynne Sheridan, the wonderful people whose judgment helps give access to the art of Bob Dylan even as they serve as judicious gatekeepers to the person of Bob Dylan.

Jessica Sindler, senior editor at Dey Street Books, first approached me about writing this book in October 2016, and I am grateful she did so, as I am for her expert help, guidance, admonishment, and encouragement.

Particular thanks are due to Jud Herrman, Tim Joseph, Kevin McGrath, and Mike Sullivan, bobfans, scholars, poets, and friends, who read a draft of the book.

Special thanks go to family, friends, and many others whose acquaintance I made over the years and who all contributed to this book: Stephen Hazan Arnoff, John "Dan" Bergan, Daniel Blanco, Barbara Boyd, Susannah Braund, Ward Briggs, John Broughton, Sergio Casali, Lydia Cawley, Michael Chaiken, Matthew Clark, James Clauss, Michael Cosmopoulos, Megan Devir, Will Dingee, Terry Gans, Elena Giusti, Joe Harris, John Henderson, Peter Knox, C. P. Lee, Catharine Mason, Tom Palaima, Hayden Pelliccia, Seth Pitman, Robert Polito, Samuel Puopolo, Iain Purdie, Sir Christopher Ricks, Chris Rollason, Ben Roy, Stephen Scobie, Jason Scorich, Lorri Shalley, Linda Stroback, Theoharis Theoharis, Joan Thomas, Julia Thomas, Sarah Thomas, Scott Warmuth, Rose Whitcomb, India Whitmarsh, Tim Whitmarsh, Elizabeth Wilson, Clem Wood, Teresa Wu, Jan Ziolkowski, Ted Ziolkowski—with apologies to those whose names should be here but aren't.

These are just some of the many people with whom I have traveled through my life with Dylan. I particularly acknowledge the dedicatees of this book, my freshman seminar students at Harvard from 2004, 2008, 2012, and 2016, teenagers who decided that Bob Dylan was worth a quarter of their time for a semester, even to the puzzlement of their peers. The announcement of October 13, 2016, was vindication for them as it was for millions of us who have in their long or less-long lives known that something was happening here.

NOTES

CHAPTER 2

23 "singing our songs": T. E. Førland, "Bringing It All Back Home," 351, also good on Dylan and the protest tradition.

24 "conscience of Young America": Bob Dylan, *Chronicles: Volume One*, 133. For the record, the degree was awarded by President Robert F. Goheen, a classics professor, as in June 2004 at St. Andrews University, where he received his other honorary doctorate from classicist and chancellor Sir Kenneth Dover.

27 tape of the show: https://www.youtube.com/watch?v=t4nA3QwGPBg&sns=em.

32 the album represented: Jonathan Cott, *Bob Dylan: The Essential Interviews*, 260.

33 *Blood on the Tracks*: Mary Travers interview, http://www.bloodonthetracks.net/About-Bob's-Album.html.

35 the album's songs: Clinton Heylin, *Behind the Shades Revisited*, 466.

CHAPTER 3

43 from Rolfzen's talk: Greil Marcus, "A Trip to Hibbing High."

44 "long-lost friend": L. Tuccio-Koonz, "Mark Twain Fan."

45 friend John Bucklen: Heylin, *Behind the Shades Revisited*, 16.

46 "run away from yourself": Ibid., 4.

48 Friends of Chile: Heylin, *Revolution in the Air*, 148–50, for an account of the song. Dylan's parents were present at the Carnegie Hall show.

50 for his last year: Heylin, *Behind the Shades Revisited*, 28.
55 *King of Kings*: Drawn from John 19:24–25.
56 the Black Sea: The subject of Chapter 8.
60 they belong together: T. E. Strunk in "Achilles in the Alleyway" even hypothesized that Dylan in the 1960s and '70s was actually reusing the poems of Catullus. While the Roman poet and the songwriter often end up in the same place, in my view that comes from a shared lyric attitude and shared themes, rather than through any intertextual nexus.
67 "open-stage policy": A. Carrera, "Oh, the Streets of Rome," 88.
68 "Girl leaves boy": Not printed in Terkel's publication of the interview, and therefore not in Cott's *Essential Interviews*. Audible at minute 33.22 of the taped version on YouTube: https://www.youtube.com/watch?v=t4nA3QwGPBg&sns=em.
68 the first time: https://www.youtube.com/watch?v=j-KEiBBg0sg.
72 end of the year: Heylin, *Revolution in the Air*, 418.
79 "pack of wild geese": Ibid., 419.
81 "every time I sing it": Cott, *The Essential Interviews*, 262.
81 antiquity of the song: Ibid., 263.

CHAPTER 4

109 "better shelves": Suze Rotolo, *A Freewheelin' Time*, 99.
112 the Greek historian (36): Thomas, "The Streets of Rome."
113 First Gulf War: Cott, *The Essential Interviews*, 385.
121 "believe in reincarnation?": Cott, *The Essential Interviews*, 194.
121 journalist Mikal Gilmore: Cott, Ibid., 411–28.
128 scripture in his life: Douglas Brinkley, "Bob Dylan's Late-Era, Old-Style American Individualism."
129 the following caption: The quote is tucked away as a caption to one of the photos included in the *AARP The Magazine.* On a table there are three old-looking well-used books, with a fourth open but not legible: http://www.aarp.org/entertainment/music/info-2015/bob-dylan-photos.html#slide12.

CHAPTER 5

131 Dylan to John Hammond: Heylin, *The Recording Sessions*, 8.
134 Cecil Sharp in 1917: Heylin, *Revolution in the Air*, 116.
134 justice of her case: Ibid., 117.
142 London at the end of 1962: Heylin, *Revolution in the Air*, 124.
146 "impersonating Bob Dylan": Sean Wilentz, *Bob Dylan in America*, 118.
149 New York in early 1963: Rotolo, *A Freewheelin' Time*, 199.
149 "'look at Rimbaud and Verlaine'": Heylin, *Behind the Shades Revisited*, 101.
153 biography Shelton was writing: Robert Shelton, *No Direction Home*, 392.
153 "writing I'm gonna do": Heylin, *Revolution in the Air*, 181.
155 "to the highest degree": Shelton, *No Direction Home*, 294.
156 as Van Ronk remembers it: Dave Van Ronk, *The Mayor of MacDougal Street*, 4.
157 politics of Dylan's art: Marqusee, *Chimes of Freedom*, 93.
158 "harvest of vitriolic verse": Paul Schmidt, *Arthur Rimbaud,* 75.

CHAPTER 6

180 finds fault with the scene: Heylin, *Revolution in the Air*, 416.
187 the new millennium: The title is stolen from Eric Lott's treatise of 1993, *Love & Theft: Blackface Minstrelsy and the American Working Class*.
188 "fatalism of classic blues": Cott, *The Essential Interviews*, 394–95.
191 Virgil and Bob Dylan both matter: Thomas, "Shadows Are Falling," for a more detailed study of the shared melancholy of Dylan and Virgil.

CHAPTER 7

199 "Lonesome Day Blues": http://dylanchords.info/41_lat/lonesome_day_blues.htm.
201 months after its release: Cott, *The Essential Interviews*, 426.
205 popularity as being: Tacitus, *Dialogue on Oratory,* 13.
206 as Suetonius reports: Suetonius, *Life of Virgil*, 11.
206 true for many songs: Ibid., 26.
211 Bob Dylan, 2004: Cott, *The Essential Interviews*, 438.
211 Bob Dylan, 1977: Heylin, *Revolution in the Air*, 181.

213 "overflow of powerful feelings": William Wordsworth, *Preface to the Lyrical Ballads*.

213 in a single session: http://www.newyorker.com/magazine/1964/10/24/the-crackin-shakin-breakin-sound.

215 after the stay in Toronto: Heylin, *Revolution in the Air*, 176–81, for good detective work on this song.

218 his own virtuoso sentence: Neil McCormick, "Bob Dylan: 30 Greatest Songs."

CHAPTER 8

233 puts it well: See Wilentz, *Bob Dylan in America*, 314–19, for an excellent detailed reading of "Nettie Moore."

241 the songs of *Modern Times*: See Thomas, "The Streets of Rome," for all the Ovidian lines; Robert Polito, "Dylan's Memory Palace," for a recent treatment of Timrod and Ovid.

241 release of *Shadows in the Night*: Robert Love, "Bob Dylan Does the American Standards His Way."

242 "from their crumbling sepulchers": Ovid, *Amores* 1.8.17; see Polito, "Dylan's Memory Palace" for other uses of Ovid's *Amores*.

243 had become part of: T. S. Eliot, "Philip Massinger."

CHAPTER 9

267 comeback album, *Time Out of Mind*: Interview with Jon Pareles in Cott, *The Essential Interviews*, 392.

268 on YouTube: https://www.youtube.com/watch?v=9UXeXdPv7rA.

269 recalls Dylan saying: Heylin, *Behind the Shades: The 20th Anniversary Edition*, 717.

270 Great American Songs: a blues: Fricke, "Bob Dylan: Shadows in the Night."

270 interview with Robert Love: Love, "Bob Dylan Does the American Standards His Way."

271 one way or another: Ibid.

274 puts together a show: Ibid.

274 "A certain kind of drama": Ibid.

275 with a perceptive observation: Heylin, *Still on the Road*, 470.

276 "where the sound's most powerful": Flanagan, "Q & A."

282 "and a sense of purpose": Edna Gunderson, "Dylan Sets the Tone for the *Wonder Boys* Soundtrack."

283 various phases of his life: Dylan addressed the issue of such change at a show at the Fox Warfield in San Francisco on November 12, 1980. In a long introduction to "Caribbean Wind" he started talking about Leadbelly: "At first he was just doing prison songs, stuff like that. The same man that recorded him also recorded Muddy Waters before Muddy Waters changed his name. He'd been out of prison for some time when he decided to do children's songs. People said all right, did Leadbelly change? Some people liked the old ones, some liked the new ones. Some people liked the prison songs and some people like the children's songs. But he didn't change, he was the same man."

283 he went off topic: Love, "Bob Dylan Does the American Standards His Way."

284 Robert Love asked him: Ibid.

CONCLUSION

293 Bob Dylan for the Nobel: Gordon Ball, "Dylan and the Nobel."

293 "Nobel Committee has gone electric": Nancy Wartick, "Readers Go Electric."

294 "an insult to real poets": *Irish Times*, October 13, 2016.

297 "matters where it takes you": Heylin, *Revolution in the Air*, 93–94. In 1965 Dylan had said, "I wrote it at the time of the Cuban crisis." Heylin pointed out that the song was first performed at Carnegie Hall on September 22, 1962, before the U-2 spy plane took photographs of missile sites in western Cuba. Dylan's original memory may not be so faulty. Cuba had been a building crisis through the summer.

299 "seen such applause before": David Gaines, "The Bob Dylan Nobel."

301 might suggest mass incarceration: 200 per 100,000 when Dylan wrote the song, now almost five times that number.

301 Dylan's Nobel Banquet Speech: Delivered by U.S. Ambassador to Sweden Azita Raji, December 10, 2016.

311 in Los Angeles on June 4: https://www.youtube.com/watch?v=6Tl cPRlau2Q

313 already saw: in Chapter 8.

320 "less and less artistic power": Michael Gray, *Song and Dance Man III*, 877.

BIBLIOGRAPHY

Ball, G. 2007. "Dylan and the Nobel." Ed. Catharine Mason and Richard Thomas. *The Performance Artistry of Bob Dylan*. *Oral Tradition* 22.1: 14–29.

Brinkley, D. 2009. "Bob Dylan's Late-Era, Old-Style American Individualism." *Rolling Stone*, May 14.

Carrera, A. 2009. "Oh, the Streets of Rome: Dylan in Italy." Ed. C. J. Sheehy and T. Swiss. *Highway 61 Revisited: Dylan's Road from Minnesota to the World*. Minneapolis: University of Minnesota Press, 84–105.

Cott, J. 2006. *Bob Dylan: The Essential Interviews*. New York: Wenner.

Dylan, B. 2004. *Chronicles: Volume One*. New York: Simon & Schuster.

Eig, J., and S. Moffett. 2003. "Did Bob Dylan Lift Lines from Dr. Saga? Don't Think Twice, It's All Right Is the View of This Japanese Writer." *Wall Street Journal*, July 8. https://www.wsj.com/articles/SB10576176194220600.

Fagles, R. 1996. Tr. Homer, *The Odyssey*. London: Penguin Classics.

Fagles, R. 2006. Tr. Virgil, *The Aeneid*. London: Penguin Classics.

Ferry, D. 1999. Tr. *Eclogues* of Virgil. New York: Farrar, Straus & Giroux.

Ferry, D. 2005. Tr. *Georgics* of Virgil. New York: Farrar, Straus & Giroux.

Flanagan, B. 2017. "Q & A with Bill Flanagan." http://bobdylan.com/news/qa-with-bill-flanagan/.

Førland, T. E. 1992. "Bringing It All Back Home *or* Another Side of Bob Dylan: Midwestern Isolationist." *Journal of American Studies* 26.3: 337–55.

Fricke, D. 2015. "Bob Dylan: Shadows in the Night." *Rolling Stone*,

February 3. http://www.rollingstone.com/music/albumreviews/bob-dylan-shadows-in-the-night-20150203.

Gaines, D. 2016. "The Bob Dylan Nobel: The Morning After." *Austin Chronicle*, December 12, 2016. http://www.austinchronicle.com/daily/arts/2016-12-12/the-bob dylan-nobel-the-morning-after/.

Giles, J. 2014. "Bob Dylan to Release an Entire Album of Frank Sinatra Songs." *Ultimate Classic Rock*, December 9. http://ultimateclassicrock.com/bob-dylan-shadows-in-the-night/?trackback=tsmclip.

Gilmore, M. 2012. "Dylan Unleashed." *Rolling Stone*, September 27. http://www.rollingstone.com/music/news/bob-dylan-unleashed-a-wild-ride-on-his-new-lp-and-striking-back-at-critics-20120927.

Gray, M. 2006. *The Bob Dylan Encyclopedia*. London: Bloomsbury.

Gray, M. 2000. *Song and Dance Man III: The Art of Bob Dylan*. London and New York: Continuum.

Green, P. 1994. Tr. Ovid, *Poems of Exile*. London: Penguin Books.

Gunderson, E. 2000. "Dylan Sets the Tone for the *Wonder Boys* Soundtrack." *USA Today*, January 20.

Hentoff, N. 1964. "The Crackin', Shakin', Breakin' Sounds." *New Yorker*, October 24. http://www.newyorker.com/magazine/1964/10/24/the-crackin-shakin breakin-sounds.

Heylin, C. 2011. *Behind the Shades: The 20th Anniversary Edition*. London: Faber & Faber.

Heylin, C. 2010. *Still on the Road: The Songs of Bob Dylan: Vol. II: 1974–2006*. Chicago: Chicago Review Press.

Heylin, C. 2009. *Revolution in the Air: The Songs of Bob Dylan: Vol. 1: 1957–73*. Chicago: Chicago Review Press.

Heylin, C. 2001. *Bob Dylan: Behind the Shades Revisited*. New York: HarperCollins.

Heylin, C. 1996. *Bob Dylan: The Recording Sessions 1960–1994*. New York: St. Martin's Griffin.

Kermode, F., ed. 1975. *Selected Prose of T. S. Eliot*. New York: Harvest Books.

Lethem, J. 2007. "The Ecstasy of Influence. A Plagiarism." *Harper's Magazine*, February. http://harpers.org/archive/2007/02/the-ecstasy-of-influence/11/.

Love, R. 2015. "Bob Dylan Does the American Standards His Way."*AARP The Magazine*, February/March. http://www.aarp.org/entertainment/style-trends/info-2015/bob-dylan-aarp-magazine.2.html.

Marcus, G. 2013. "A Trip to Hibbing High." *Riggio Honors Program: Profiles and Reflections*, October. http://riggio.americanvanguardpress.com/portfolio/a-trip-to-hibbing-high-greil-marcus/.

Marqusee, M. 2003. *Chimes of Freedom. The Politics of Bob Dylan's Art*. New York: New Press.

McCormick, N. 2013. "Bob Dylan: 30 Greatest Songs." *Telegraph*, November 18.

Polito, R. 2013. "Dylan's Memory Palace." *Riggio Honors Program: Profiles and Reflections*, July. http://riggio.americanvanguardpress.com/portfolio/bob dylans-memory-palace-robert-polito/.

Ricks, C., and J. McCue, eds. 2015. *The Poems of T. S. Eliot*. Vol. 1. Baltimore: Johns Hopkins University Press.

Rogovoy, S. 2009. *Bob Dylan: Prophet, Mystic, Poet*. New York: Scribner.

Rotolo, S. 2008. *A Freewheelin' Time: A Memoir of Greenwich Village in the Sixties*. New York: Broadway Books.

Santelli, R., and B. Dylan. 2005. *The Bob Dylan Scrapbook, 1956–1966*. New York: Simon & Schuster.

Schmidt, P. 1976. *Rimbaud. Complete Works*. New York: Harper & Row.

Shawcross, J. T., ed. 1971. *The Complete Poetry of John Milton*. New York: Anchor Books.

Shelton, R. 1986. *No Direction Home. The Life and Music of Bob Dylan*. New York: Ballantine Books.

Sótano Beat. 2016. "Patti Smith—A Hard Rain's A-Gonna Fall (ceremonia Nobel 2016)." *YouTube*, December 10. https://www.youtube.com/watch?v=DVXQaOhpfJU.

Stevenson, S. 2004. "Tangled Up in Boobs." *Slate*, April 12. http://www.slate.com/articles/business/ad_report_card/2004/04/tangled_up_in_boobs.html.

Strunk, T. E. 2009. "Achilles in the Alleyway: Bob Dylan and Classical Poetry and Myth." *Arion: A Journal of Humanities and the Classics* 17.1: 119–36.

Theoharis, T. C. 2001. Tr. *Before Time Could Change Them. The Complete Poems of Constantine P. Cavafy*. New York: Harcourt.

Thomas, R. F. 2006a. "The Streets of Rome: The Classical Dylan." In C. Mason and R. F. Thomas, eds. *The Performance Artistry of Bob Dylan*, *Oral Tradition* 22.1: 30–56. http://journal.oraltradition.org/files/articles/22i/Thomas.pdf.

Thomas, R. F. 2006b. "Shadows Are Falling: Virgil, Radnóti, and Dylan, and the Aesthetics of Pastoral Melancholy." *Rethymnon Classical Studies* 3: 191–214.

Tuccio-Koonz, L. 2016. "Mark Twain Fan Visits His Hartford Mansion, Finds It's Like Communing with a Long-Lost Friend." *CTPost*, October 30. http://www.ctpost.com/living/article/Mark-Twain-fan-visits-his-Hartford-mansion-finds-10421133.php.

Van Ronk, D. 2006. *The Mayor of MacDougal Street: A Memoir*. Boston: Da Capo Press.

Wartick, N. 2016. "Readers Go Electric Over Dylan's Nobel." *New York Times,* October 14. https://www.nytimes.com/2016/10/15/arts/music/readers-go-electric-over-dylans-nobel.html?_r=0?.

Wilentz, S. 2010. *Bob Dylan in America.* New York: Anchor Books.

BOB DYLAN LYRICS—COPYRIGHT INFORMATION

Chimes of Freedom. © Special Rider Music 1992. All rights reserved. International copyright secured. Reprinted by permission.

Cold Irons Bound. © Special Rider Music 1997. All rights reserved. International copyright secured. Reprinted by permission.

Day of the Locusts. © Big Sky Music 1998. All rights reserved. International copyright secured. Reprinted by permission.

Desolation Row. © Special Rider Music 1993. All rights reserved. International copyright secured. Reprinted by permission.

Dirt Road Blues. © Special Rider Music 1997. All rights reserved. International copyright secured. Reprinted by permission.

Don't Think Twice, It's All Right. © Special Rider Music 1991. All rights reserved. International copyright secured. Reprinted by permission.

Duquesne Whistle. © Special Rider Music and Ice Nine Publishing 2012. All rights reserved. International copyright secured. Reprinted by permission.

Early Roman Kings. © Special Rider Music 2012. All rights reserved. International copyright secured. Reprinted by permission.

Floater (Too Much to Ask). © Special Rider Music 2001. All rights reserved. International copyright secured. Reprinted by permission.

Forgetful Heart. © Special Rider Music and Ice Nine Publishing 2009. All rights reserved. International copyright secured. Reprinted by permission.

Fourth Time Around. © Dwarf Music 1994. All rights reserved. International copyright secured. Reprinted by permission.

Girl from the North Country. © Special Rider Music 1991. All rights reserved. International copyright secured. Reprinted by permission.

Gonna Change My Way of Thinking. © Special Rider Music 1979. All rights reserved. International copyright secured. Reprinted by permission.

High Water (For Charley Patton). © Special Rider Music 2001. All rights reserved. International copyright secured. Reprinted by permission.

Highlands. © Special Rider Music 1997. All rights reserved. International copyright secured. Reprinted by permission.

Honest with Me. © Special Rider Music 2001. All rights reserved. International copyright secured. Reprinted by permission.

Huck's Tune. © Special Rider Music 2007. All rights reserved. International copyright secured. Reprinted by permission.

I Shall Be Free No. 10. © Special Rider Music 1999. All rights reserved. International copyright secured. Reprinted by permission.

I'll Remember You. © Special Rider Music 1985. All rights reserved. International copyright secured. Reprinted by permission.

Idiot Wind. © Ram's Horn Music 2002. All rights reserved. International copyright secured. Reprinted by permission.

Is Your Love in Vain? © Special Rider Music 1978. All rights reserved. International copyright secured. Reprinted by permission.

Isis. © Special Rider Music 2003. All rights reserved. International copyright secured. Reprinted by permission.

It Ain't Me Babe. © Special Rider Music 1992. All rights reserved. International copyright secured. Reprinted by permission.

Just Like Tom Thumb's Blues. © Special Rider Music 1993. All rights reserved. International copyright secured. Reprinted by permission.

Kingsport Town. © Special Rider Music 1991. All rights reserved. International copyright secured. Reprinted by permission.

Lonesome Day Blues. © Special Rider Music 2001. All rights reserved. International copyright secured. Reprinted by permission.

Long Ago, Far Away. © Special Rider Music 1996. All rights reserved. International copyright secured. Reprinted by permission.

Love Sick. © Special Rider Music 1997. All rights reserved. International copyright secured. Reprinted by permission.

Make You Feel My Love. © Special Rider Music 1997. All rights reserved. International copyright secured. Reprinted by permission.

Million Miles. © Special Rider Music 1997. All rights reserved. International copyright secured. Reprinted by permission.

Mississippi. © Special Rider Music 1997. All rights reserved. International copyright secured. Reprinted by permission.

Most of the Time. © Special Rider Music 1989. All rights reserved. International copyright secured. Reprinted by permission.

Mr. Tambourine Man. © Special Rider Music 1993. All rights reserved. International copyright secured. Reprinted by permission.

My Wife's Home Town. © Special Rider Music, Ice Nine Publishing and Hoochie Coochie Music 2012. All rights reserved. International copyright secured. Reprinted by permission.

Narrow Way. © Special Rider Music 2012. All rights reserved. International copyright secured. Reprinted by permission.

Nettie Moore. © Special Rider Music 2008. All rights reserved. International copyright secured. Reprinted by permission.

Never Say Goodbye. © Ram's Horn Music 2001. All rights reserved. International copyright secured. Reprinted by permission.

Not Dark Yet. © Special Rider Music 1997. All rights reserved. International copyright secured. Reprinted by permission.

Pay in Blood. © Special Rider Music 2012. All rights reserved. International copyright secured. Reprinted by permission.

Po' Boy. © Special Rider Music 2001. All rights reserved. International copyright secured. Reprinted by permission.

Positively 4th Street. © Special Rider Music 1993. All rights reserved. International copyright secured. Reprinted by permission.

Rollin' and Tumblin'. © Special Rider Music 2006. All rights reserved. International copyright secured. Reprinted by permission.

Sara. © Ram's Horn Music 2004. All rights reserved. International copyright secured. Reprinted by permission.

Scarlet Town. © Special Rider Music 2012. All rights reserved. International copyright secured. Reprinted by permission.

Someday Baby. © Special Rider Music 2006. All rights reserved. International copyright secured. Reprinted by permission.

Spirit on the Water. © Special Rider Music 2006. All rights reserved. International copyright secured. Reprinted by permission.

Standing in the Doorway. © Special Rider Music 1997. All rights reserved. International copyright secured. Reprinted by permission.

Tangled Up in Blue. © Ram's Horn Music 2002. All rights reserved. International copyright secured. Reprinted by permission.

Tell Me That It Isn't True. © Big Sky Music 1997. All rights reserved. International copyright secured. Reprinted by permission.

Tempest. © Special Rider Music 2012. All rights reserved. International copyright secured. Reprinted by permission.

Temporary Like Achilles. © Dwarf Music 1994. All rights reserved. International copyright secured. Reprinted by permission.

The Groom's Still Waiting at the Altar. © Special Rider Music 1981. All rights reserved. International copyright secured. Reprinted by permission.

The Levee's Gonna Break. © Special Rider Music 2006. All rights reserved. International copyright secured. Reprinted by permission.

Things Have Changed. © Special Rider Music 1999. All rights reserved. International copyright secured. Reprinted by permission.

This Dream of You. © Special Rider Music 2009. All rights reserved. International copyright secured. Reprinted by permission.

Thunder on the Mountain. © Special Rider Music 2006. All rights reserved. International copyright secured. Reprinted by permission.

'Til I Fell in Love with You. © Special Rider Music 1997. All rights reserved. International copyright secured. Reprinted by permission.

Tin Angel. © Special Rider Music 2012. All rights reserved. International copyright secured. Reprinted by permission.

Tryin' to Get to Heaven. © Special Rider Music 1997. All rights reserved. International copyright secured. Reprinted by permission.

When I Paint My Masterpiece. © Big Sky Music 1999. All rights reserved. International copyright secured. Reprinted by permission.

When the Deal Goes Down. © Special Rider Music 2006. All rights reserved. International copyright secured. Reprinted by permission.

Why Try to Change Me Now. © Notable 2015. All rights reserved. International copyright secured. Reprinted by permission.

Workingman's Blues #2. © Special Rider Music 2006. All rights reserved. International copyright secured. Reprinted by permission.

You're a Big Girl Now. © Ram's Horn Music 2002. All rights reserved. International copyright secured. Reprinted by permission.

You're Gonna Make Me Lonesome When You Go. © Ram's Horn Music 2002. All rights reserved. International copyright secured. Reprinted by permission.

INDEX